The Purpose-Driven Organisation in Times of Climate Crisis

This book provides an authoritative, evidence-based understanding of the implications of climate change for organisations and offers valuable insights into how to craft and embed a sustainable purpose.

The book draws upon the newly emerging cadre of 'purpose professionals' who are charged with shifting their organisation from being shareholder-driven to stakeholder-driven through the development and implementation of organisational purpose. It represents a unique collaboration between world-leading climate scientists and management experts, explaining in clear terms how organisations have been implicated in climate change, and the consequences of failing to act now. Based on science, it sets out the case for organisations to effect a radical shift away from a shareholder imperative to become purpose-driven, and provides actionable, evidence-based insights into how to achieve change successfully. In the first chapters, climate scientists outline the key scientific evidence about how the Earth's climate has changed in the past 100 years, and use case studies to explain how this will affect policy, the public, and organisations across different geographical regions and sectors. Next, organisational experts explain the regulatory context and policy drivers for the shift to organisational purpose. We then focus on organisational actions and priorities, drawing on latest research findings to explain what purpose is, and how to lead change effectively to overcome the obstacles to becoming purpose-led.

Written with a practitioner readership in mind, this book is an invaluable resource for senior leaders, corporate social responsibility practitioners, policymakers, and business and management students studying modules in purpose-driven leadership and climate change as well as a wider lay readership.

Katie Bailey is Professor of Leadership and HRM at Northumbria University and Emerita Professor of Work and Employment at King's College London. She is a widely published, international award-winning scholar, focusing on meaningful work, employee engagement, temporality at work, and human resource management. She was the lead editor of *Employee Engagement in Theory and Practice* (2013, Routledge) and co-editor of the *Oxford Handbook of Meaningful Work* (2019) as well as the lead author of *Strategic Human Resource Management* (2nd Ed., 2018).

Katie Manning is Lecturer in Climate Change, Business, and Society in the Department of Geography at King's College London, UK. Her work focuses on cross-disciplinary, and systems approaches to climate change, drawing on the political, economic, and social dimensions of climate mitigation and adaptation. Katie received her PhD from the University of Oxford, looking at the role of human adaptation to past climatic change and, having worked in the field of palaeoclimatology for over a decade, has more recently shifted her attention to the challenge of contemporary and future climate change.

Routledge Research in Sustainability and Business

Circular Economy in the European Union
Organisational Practice and Future Directions in Germany, Poland and Spain
Dagmara Lewicka, Joanna Zarębska, Roman Batko, Maciej Woźniak, Dariusz Cichoń, Beata Tarczydło, Monika Pec

Degrowth, Depth and Hope in Sustainable Business
Reflections from Denmark, Finland and Sweden
Iana Nesterova

Green Innovation in Central and Eastern Europe
Iryna Bashynska, Dagmara Lewicka, Svitlana Filyppova and Olha Prokopenko

Sustainability Reporting in Central and Eastern Europe Countries
Marzena Remlein and Ana Rep

Explainable Artificial Intelligence for Sustainable Development
Advancing Social and Economic Transformations
Edited by Ewa Wanda Ziemba, Wioletta Grzenda and Michal Ramsza

Integrating Sustainability and ESG in Private Equity
A Playbook for Generating Long-Term Value from Sustainable Practices
Esohe Denise Odaro and Sasja Beslik

The Purpose-Driven Organisation in Times of Climate Crisis
Edited by Katie Bailey and Katie Manning

For more information about this series, please visit: www.routledge.com/Routledge-Research-in-Sustainability-and-Business/book-series/RRSB

The Purpose-Driven Organisation in Times of Climate Crisis

Edited by Katie Bailey
and Katie Manning

Routledge
Taylor & Francis Group
LONDON AND NEW YORK

Designed cover image: Getty Images

First published 2026
by Routledge
4 Park Square, Milton Park, Abingdon, Oxon OX14 4RN

and by Routledge
605 Third Avenue, New York, NY 10158

Routledge is an imprint of the Taylor & Francis Group, an informa business

British Library Cataloguing-in-Publication Data
A catalogue record for this book is available from the British Library

ISBN: 978-1-032-89896-4 (hbk)
ISBN: 978-1-032-89895-7 (pbk)
ISBN: 978-1-003-54522-4 (ebk)

DOI: 10.4324/9781003545224

Typeset in Times New Roman
by Apex CoVantage, LLC

Katie Bailey would like to dedicate this book to her grandson Harry, in the hope that our generation will leave behind a healthy, sustainable planet where his generation can flourish.

Katie Manning would like to dedicate this book to her children, Leon and Fisher, and to the next generation of students, practitioners, and policymakers striving to build a more equitable and climate-resilient future.

Contents

List of figures *xi*
List of tables *xii*
List of boxes *xiii*
List of contributors *xv*
Acknowledgements *xviii*

Introduction 1
KATIE BAILEY AND KATIE MANNING

1 **The science of human-induced climate change** 24
MARK MASLIN

2 **The impacts of global climate change in the context of organisational responsibility** 49
ANITA WREFORD AND KATIE MANNING

3 **The global context for organisational climate action** 75
DUNCAN RUSSEL AND NICK KIRSOP-TAYLOR

4 **The shift from shareholder capitalism to marginalised stakeholder capitalism** 91
RASHEDUR CHOWDHURY

5 **Defining and developing organisational purpose** 110
VICTORIA HURTH AND CHARLES EBERT

6 **Effecting the change to become purpose-led** 134
CRAIG V. ROBINSON AND LUIS D. TORRES

7 **Purpose-led decision-making across the organisation** 152
CATHERINE TILLEY

8 **Managing people in the purpose-led organisation** 170
NA FU

9 **Purpose-led communication: practical steps towards defining your organisation's purpose and communicating meaningfully to your stakeholders** 187
JOEP CORNELISSEN AND MIRJAM WERNER

10 **Measuring and reporting on purpose at a time of crisis and unravelling consensus** 201
DAVID AIKMAN AND MARC LEPERE

Index *227*

Figures

0.1 Timeline showing the key scientific, governance, and business/financial developments related to climate change with respect to temperature records since 1850 and carbon dioxide (CO_2) records since 1958. 4
1.1 Explanation of the greenhouse effect. 25
1.2 Greenhouse gases and temperature for the last eight glacial cycles recorded in ice cores. 26
1.3 Mauna Loa observatory atmospheric carbon dioxide (CO_2) measurements. 28
1.4 The carbon cycle showing the main carbon reservoirs in the oceans and land biosphere. 38
1.5 Future carbon emission scenarios and potential global warming by 2100. 42
2.1 Environmental Kuznets curve, showing the relationship between greenhouse gas emissions and economic growth and potential for decoupling. 51
2.2 Behavioural drivers and feedback loops between climate change, socio-economic impacts, and organisational/individual action. 53
7.1 Contextual layers influencing decision-making. 156
7.2 Organisational factors influencing decision-making. 167
8.1 The employee life cycle. 172
9.1 Starting with purpose: The communication model canvas©. 189
9.2 Three parts of the overall purpose narrative. 198

Tables

1.1	Main greenhouse gases and their comparative ability to warm the atmosphere.	36
1.2	Shared socio-economic pathways representing projected socio-economic global changes up to 2100.	41
1.3	Combinations of RCPs and SSPs showing the five main scenarios that were the focus in the latest IPCC Assessment Report (AR6).	41
2.1	Sector contributions to GHG emissions showing the annual contribution of total anthropogenic emissions.	52
2.2	Summary of potential climate impacts on infrastructure.	64
4.1	Marginalised Stakeholder capabilities.	102
5.1	ISO37000 factors for determining an organisational purpose and reflection questions.	123
10.1	Summary of key climate and purpose disclosure frameworks.	205
10.2	Estimated impact of a 10% rise in financed GHG emissions for cost of equity.	221
10.3	Eight decision-critical questions for corporate leaders to consider.	223

Boxes

2.1	Australia's Great Barrier Reef	54
2.2	Organisational actions to avert biodiversity loss	56
2.3	Are we witnessing the death of Champagne?	57
2.4	Organisational actions to mitigate food security impacts	58
2.5	Heat-related impacts on productivity	59
2.6	Organisational actions to mitigate climate change impacts on human health	60
2.7	Organisational actions to mitigate climate change-induced conflict	62
2.8	Organisational actions to mitigate climate change impacts on infrastructure and financial risk	66
3.1	Major UNFCCC actions and agreements	77
3.2	The failure of COP15	79
3.3	Climate finance under the UNFCCC	81
3.4	The most prominent negotiating groups in UNFCCC negotiations	83
3.5	The UNFCCC coordination platforms, which facilitate global climate action through collaboration, finance, and knowledge sharing	86
4.1	Asia Energy	94
4.2	Shell in Nigeria	95
4.3	Rana Plaza	99
6.1	The purpose-driven leader's dilemma	136
6.2	A renewed commitment to sustainable finance	137
6.3	The five stages of organisational preparedness	140
6.4	Cisco's approach to culture and purpose	145
8.1	Patagonia	174
8.2	Unilever	175
8.3	Danone	176
8.4	Salesforce	178
8.5	Deloitte	179
9.1	Diagnostic questions for formulating a purpose communication strategy	191
9.2	Purpose statements—Best practice guidelines	193

9.3	Activating the organisation's purpose through a purpose narrative	196
10.1	Reducing food waste: Essendi	215
10.2	Optimising warehousing for sustainability and efficiency: SH LLC	217

Contributors

David Aikman is Director of the National Institute for Economic and Social Research and was formerly Professor of Practice in Finance and Director of the Qatar Centre for Global Banking and Finance at King's College London. He spent 17 years as an economist at the Bank of England and currently advises various central banks on monetary policy and financial stability policy issues.

Katie Bailey is Professor of Leadership and HRM at Northumbria University and an Emerita Professor of Work and Employment at King's College London. Her research focuses on purpose and meaningful work. Among her many award-winning publications, she is also co-editor of the *Oxford Handbook of Meaningful Work*.

Rashedur Chowdhury is Professor of Business and Management at Essex Business School, University of Essex, and a Batten Fellow at the Darden School of Business, University of Virginia. His recent works focus on the Rana Plaza collapse and the Rohingya refugee crisis in Bangladesh.

Joep Cornelissen is Professor of Corporate Communication and Management at Rotterdam School of Management, Erasmus University, and Chair of Strategy and Organization (part-time) at the University of Liverpool Management School. He studies the role of communication in the context of innovation, entrepreneurship, and change.

Charles Ebert is a graduate from Cambridge University, where he completed his PhD on Organisational Purpose. In addition to his current role as Assistant Professor at the University of Wisconsin Eau Claire, Dr. Ebert also serves as an external advisor for Contexis, a consultancy that specialises in purpose-driven business.

Na Fu is Professor of Responsible Leadership and Fellow at Trinity College Dublin. Her research focuses on responsible leadership, employee well-being, and digital transformation. With over €7.9 million in research funding, she leads and contributes to major EU projects, shaping digital skills policy and fostering inclusive, future-ready workplaces.

Victoria Hurth is an independent pracademic working to align humanity's decision-making with collective long-term well-being. She is Fellow of the

University of Cambridge Institute for Sustainability Leadership, leads ISO standard development in purpose governance, and is an executive and impact advisor as well as an academic lead for Common Ground Research Networks.

Nick Kirsop-Taylor is an environmental governance scholar at the University of Essex, focused on how the state and bureaucracy contribute towards and participate in contestation and conflict in environmental governance. He has co-authored the *Handbook of the Governance of Sustainable Development* and *Governance in the Age of Nature Restoration*.

Marc Lepere is Professor of Practice in Sustainability at King's College London. He was Chief Marketing Officer for Havas, Executive Vice President for McCann Erickson, and Deputy Chairman of Dentsu. He has published in numerous outlets, including a Pulitzer-prize-nominated article in *Discovering Truth.*

Katie Manning is Systems Scientist at King's College London, whose research spans past, contemporary, and future response to climatic change. Her work addresses interdependencies between socio-economic, environmental, and political dimensions of climate change, focusing on nature-based solutions for climate adaptation and mitigation.

Mark Maslin is Professor of Earth System Science at UCL. He is joint Pro-Vice-Provost of the UCL Climate Crisis Grand Challenge and was Specialist Advisor for the UK Parliament Environment Audits Committee on climate change and security. In 2023, Onalytica ranked him #1 Global Sustainability thought leader and influencer.

Craig V. Robinson is Professor in Business Education at King's College London. Craig teaches and supervises action research in strategy, change management, and leadership and has worked with executive students in the UK, Europe, the Middle East, Asia, and Africa since starting in academia in 2007.

Duncan Russel is Professor of Environmental Policy with a focus on biodiversity and climate policy in different governance settings. He has published many works, including the *Edward Elgar Handbook of the Governance of Sustainable Development* (2022).

Catherine Tilley is Reader in Business Ethics and Sustainability and Impact Director at the Centre for Sustainable Business at King's College London. Previously, she worked in management consultancy for over 20 years, focusing on strategic and organisational issues.

Luis D. Torres is Associate Professor and the Director of Executive Education at Nottingham University Business School (UK), Associate Professor at the Universidad del Desarrollo in Chile, and a Data Intelligence Consultant for the United Nations High Commissioner for Refugees. His expertise lies in CSR/ESG, regulation, and data intelligence.

Mirjam Werner is Associate Professor in the Business-Society Management Department at Rotterdam School of Management, Erasmus University. She studies the ways in which individuals and organisations challenge the status quo and, through entrepreneurship and campaigning, fight for a more equitable and sustainable world.

Anita Wreford is Professor in the field of applied economics at Lincoln University, New Zealand, specialising in adaptation to climate change. She is the lead author on two IPCC reports and provides advice on climate adaptation for the EU Commission, the OECD, Scottish Government, and the UK's Climate Change Committee.

Acknowledgements

Katie Bailey would like to extend her heartfelt thanks to members of the Meaning and Purpose Network based at King's College London, and organisers and members of the Purpose-in-Practice Community for the spirited, engaging, and challenging conversations about purpose which inspired this book.

Katie Manning would like to express her gratitude to colleagues in the Department of Geography at King's College London for their support and inspiration. She would also like to extend her thanks to all the farmers and organisations working tirelessly to secure a healthy and sustainable food system, who have so generously given their time to share their wisdom and knowledge.

Chapter 7: Catherine Tilley records her gratitude to the 40 anonymous interviewees who were so generous in sharing their knowledge and experience.

Introduction

Katie Bailey and Katie Manning

On the morning of Tuesday 7 January 2025, Lucy Sherriff watched from the window of her Palisades Village home on Sunset Boulevard in Los Angeles as small flames appeared in the distance on the Santa Monica mountainside. Within an hour, the fire was making its way down the hill, whipped up by strong winds, quickly spreading across the dry undergrowth and swallowing up houses as it went. Gathering together family paperwork, she joined the long queue of people following the mandatory evacuation order that by now covered the entire Palisades district. By Wednesday morning, her home, along with 90% of other buildings in the area, including houses, schools, shops, and community centres, were all burnt to the ground (Sherriff, 2025).

The fires that raged across Los Angeles claimed at least 25 lives and destroyed over 20,000 structures. They impacted nearly 2,000 businesses, left tens of thousands of homes without power, and emitted clouds of noxious smoke into the atmosphere. Many years of drought in the region had been followed by excessive rainfall during 2022–2023, which encouraged the rapid growth of vegetation, only to be followed by further droughts in 2024, creating a tinderbox. The US National Oceanic and Atmospheric Administration said, 'climate change, including increased heat, extended drought, and a thirsty atmosphere, has been a key driver in increasing the risk and extent of wildfires in the western United States' (Fitzgerald, 2025; World Meteorological Organization, 2025).

Ten thousand miles away, Charles Eris is a Senior Livelihoods Officer at the charity Concern Kenya. He grew up in a nomadic community in Turkana in northern Kenya that depends on livestock, such as cattle, goats, and sheep, both for food and for their livelihood. Every year, the amount of rain that falls in the region has decreased, and the community's social and economic systems have effectively broken down. People are struggling to feed their families and, as well as suffering from malnutrition, are inevitably also experiencing depression and anxiety in the face of what seems like a hopeless future. The daily walk to find water is over 11km, and the situation is getting worse each year. Climate change is pushing people in the region into destitution—the current Global Acute Malnutrition rate stands at 26.4%, well in excess of the emergency threshold of 15% (Concern UK, 2024).

Over in the UK, the links between human-driven climate change and biodiversity crises are all too evident. The most recent UK Climate Risk Assessment

DOI: 10.4324/9781003545224-1

found that the average surface temperature in the UK has risen by 1.2°C since pre-industrial times (UK Government, 2022). One study has revealed that as many as 97% of wildflower meadows and 80% of chalk grasslands have been eradicated since the 1940s. The researchers conclude, 'the coming two decades will see changes in conditions of an approximately equivalent magnitude to the entire period between the present and the early 20th century' (Wilson & Pescott, 2023, p. 2002). As a result, habitats including peatland bogs that have already been significantly reduced will be most affected. Peatlands are a vital component in the ecosystem, since they absorb huge amounts of carbon, and their loss has a disproportionate effect on global warming. Alongside this, there will be radical changes to the kind of crops that can be grown, together with increased risks of damage caused by extreme weather events (Wilson & Pescott, 2023).

But, whilst countries such as the UK are facing challenges to their water resources and agricultural systems, Small Island Developing States (SIDS) are facing a much greater existential crisis. A government task force in the south Pacific country of Fiji, which comprises more than 300 low-lying islands, is busy developing plans for relocating 42 villages within the next five to ten years in the face of rising sea levels and the increasing risk of cyclones. Six villages have already been moved. The challenges of doing this are immense—it is not just houses that have to be moved but also community buildings, schools, health centres, roads, and even burial grounds. Many things need to be considered, including the availability of suitable land, and much can go wrong (Lyons, 2022). Some SIDS, such as Tuvalu, Suriname, and the Cook Islands, are increasingly exploring digital nationality to safeguard their culture and sovereignty in the face of displacement due to sea level rise. These problems are likely to be experienced by increasing numbers of people around the world over the coming years. The oceans have risen between 20 and 23 centimetres since 1880, but by 2023, had reached a record high. The rate of increase over the last ten years is more than twice that recorded between 1993 and 2002, caused by ocean warming and the melting of glaciers and ice sheets which are driven by climate change. The 900 million people around the world who live near the sea will all be affected, including those in countries such as Bangladesh, China, India, the Netherlands, and Pakistan, which will face potentially catastrophic flooding. Major cities on every continent are at risk, including Bangkok, Buenos Aires, Lagos, London, Mumbai, New York, and Shanghai (Dickinson, 2024). Where will all these people go?

Wherever we turn, the effects of human-driven—or anthropogenic—climate change prompted by the burning of fossil fuels, along with the excessive use of scarce planetary resources, land use change, and deforestation, are affecting people, animals, and the environment, as well as businesses and organisations. Unless humanity makes drastic changes, these impacts will escalate, threatening the sustainability of life on Earth in its current form—not in some distant future, but within the lifetime of our children and grandchildren. Our generation has a responsibility to ensure we leave behind a healthy, habitable planet and a bright future for the coming generations to enjoy. Are we prepared to take the actions needed to turn this into a reality?

There is widespread agreement that the changes required are deep, pervasive, and systemic (Cloutier et al., 2024). Governments have a responsibility to set the policy and regulatory context for new ways of living and new priorities (Mayer, 2021), and we as individuals need to re-evaluate the sustainability of our behaviours and lifestyles. However, environmental collapse can only be averted if firms stop focusing on profit as the only valid criterion of success and start developing and embedding a broader purpose that puts the long-term well-being of people and planet first. Unfortunately, the evidence suggests that thus far they are failing to do so: fossil fuels still account for 82% of the world's energy consumption (Nyberg & Wright, 2024), and greenhouse gas emissions from the burning of fossil fuels have increased by over 60% since 1990 (Statista, 2025).

Figure 0.1 demonstrates the gradual development of the global response to climate change, starting with the very first recognition of the link between greenhouse gas emissions and global warming in the late 19th century up to the latest financial and organisational standards for climate action. Each of these developments is discussed in detail throughout this book, highlighting the concomitant rise in climate change impacts, governance responses, and business adaptation. However, as shown in the temperature and carbon dioxide (CO_2) trends, the last 100 years of climate action have done nothing to abate the upward curve in emissions and global warming, demonstrating the critical need for transformative change in the way organisations respond to the growing climate crisis.

This book was inspired by conversations that we, the two editors of this book (one a social scientist and the other a climate scientist), have held over a number of years with senior leaders and 'purpose professionals'—people tasked by their employer with driving forward a purpose-led approach within their organisation. While some of them worked for organisations specifically set up with a pro-social purpose, most did not—they worked for large, established organisations that were trying to shift course and retrofit purpose into a setting designed, run, and regulated to maximise profits. Perhaps unsurprisingly, they told us of the many challenges they faced along the way, such as determining an appropriate purpose and knowing how to go about embedding it across the organisation. In particular, they wanted to know:

- What is the research evidence base about how climate change is affected by, and affects organisations?
- How can we persuade the sceptics within the organisation that we need to become purpose-led?
- How do we develop a purpose and integrate it into core decision-making processes?
- How can we resist the call back to 'business-as-usual'?
- How do we measure progress and maintain momentum?

The people we spoke to wanted to hear about the latest thinking and evidence relating to purpose from reliable sources to help them in their task. In an era of misinformation and fake news, it can be hard to know who to trust. Academic research

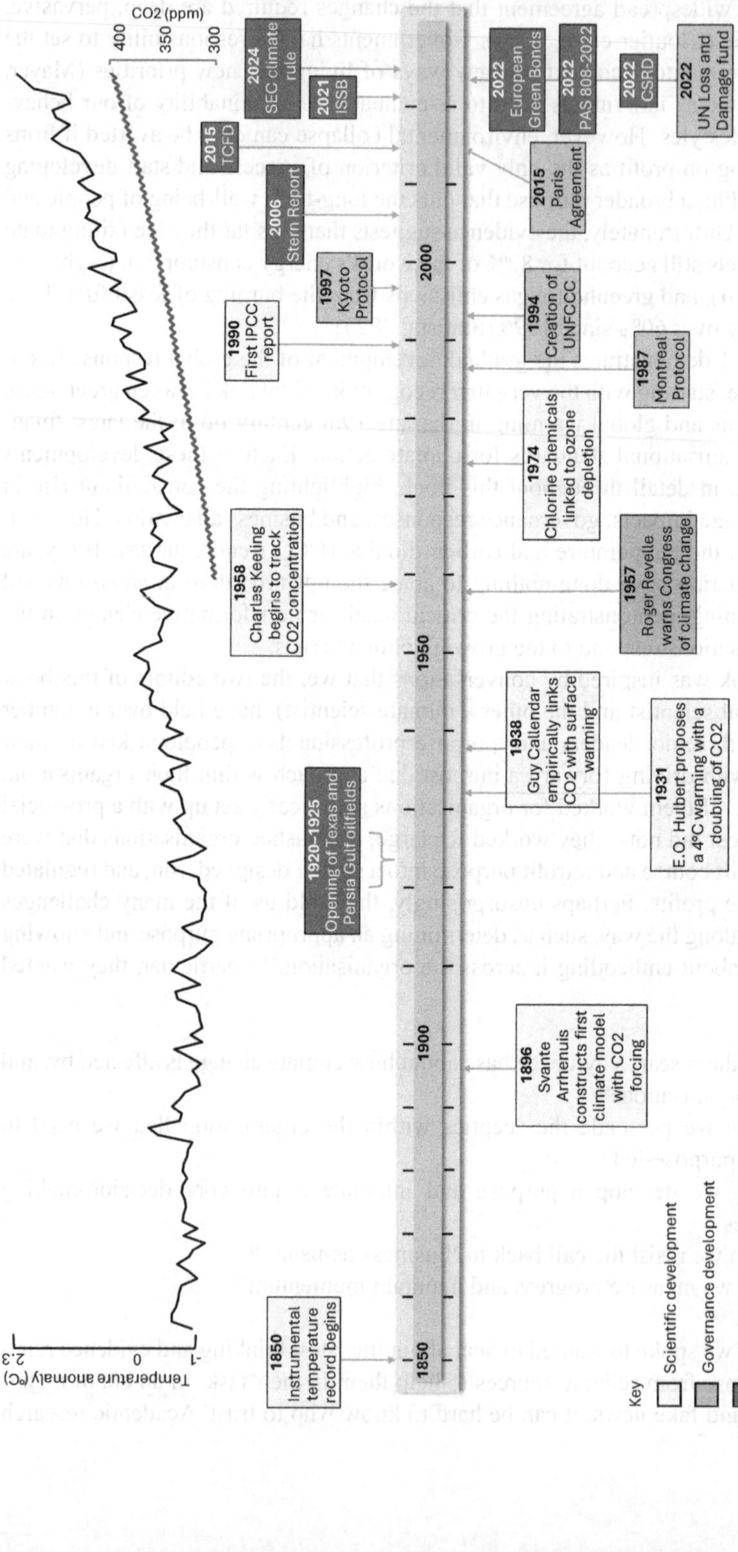

Figure 0.1 Timeline showing the key scientific governance, and business/financial developments related to climate change with respect to temperature records since 1850 and carbon dioxide (CO_2) records since 1958.

is generally regarded as more trustworthy as it has been peer-reviewed, but can be overwhelming in volume and complexity. It is often inaccessible to a practitioner readership—hidden behind a paywall or written in opaque language.

We saw an urgent need to bridge this gap. There are many scholars doing important research on all aspects of purpose that would be of great use and value to practitioners dealing with these issues, and we wanted to make this evidence available to them. We knew that we would need to invite experts across multiple disciplines to contribute—from climate scientists and policy experts to management, communication, sustainability, and finance scholars. We are delighted that so many world-leading thinkers in their fields accepted our invitation and rose to the challenge of distilling many years of research into a chapter that summarises the evidence and makes practical recommendations for action.

In this Introduction, we make the case for why organisations should be transforming to become purpose-driven; we explore what purpose means, and we consider why it is so difficult for organisations to effect the change. But first, we look in more detail at the causes and consequences of climate change.

The climate crisis: What is the evidence?

Addressing climate change is one of the most urgent priorities for society today. Over the last 175 years, greenhouse gas concentrations have risen exponentially, leading to rapid global warming, sea level rise, and more extreme weather events. These changes in the Earth system pose a significant threat to various aspects of life, including human health, ecosystem functioning, food and water security, infrastructural resilience, and sociopolitical stability. The year 2024 was the world's hottest on record and was the first calendar year to surpass 1.5°C of warming compared to pre-industrial levels. Whilst our climate has changed throughout history, the rapid warming observed over the last century is exceptional, and can be directly attributed to human activity, principally the burning of Earth's fossil fuels.

The environment we live in today, with its lush vegetation and spectacular diversity of plants and animals, is in existence because of the carbon cycle. Forged in the heart of ageing stars, carbon is the fourth most abundant element in the Universe and supports all life on Earth. Yet, the balance of the Earth's carbon cycle, the process in which carbon flows between the atmosphere, land, and ocean, is also tightly interwoven with global climate change.

Most carbon is stored in rocks and sediments, while the rest is stored in the ocean, atmosphere, and living organisms. These are the reservoirs, or sinks, through which carbon cycles. Yet, small perturbations in the amount of carbon in one reservoir can cause a ripple effect through the others. This feedback process is what triggers the Ice Ages as cyclical variations in the Earth's orbit alter the amount of energy that Earth receives from the Sun, resulting in changes in the amount of atmospheric carbon. But, whilst the waxing and waning of the glacial cycle is driven by external changes in orbital configuration, the recent increase in CO_2 levels and other greenhouse gases (GHGs), such as methane (CH_4) and nitrous oxide (N_2O), cannot be explained by exogenous forces.

The orbital cycles that create ice ages happen over a period of roughly 100,000 years, yet the increase in GHGs since the industrial revolution has occurred in less than 200 years. Carbon dioxide levels during the last glacial maximum, which occurred between 26,000 and 20,000 years ago, were about 180 parts per million (ppm), whereas pre-industrial levels from 1850 were at about 280 ppm, representing a 55% increase over that time span. Today, CO_2 is over 427 ppm, representing another 50% increase, but in just 175 years. Whilst these numbers may appear tiny, with the main GHGs accounting for just over 0.04% of molecules in the air, they are responsible for the Earth's Greenhouse effect, warming the effective temperature by 33°C. Without these critical gases, water would be frozen and the mean temperature on the planet would drop to around −18°C, making it uninhabitable.

So, how have humans perturbed the carbon cycle? And, how do we know that recent global warming is human-induced? Fossil fuels, including coal, oil, and natural gas, are formed from the ancient deposition of dead plants and animals, which are buried and subjected to intense heat and pressure, locking carbon into the Earth's rocks and sediments. When these fuels are burned, that carbon is released as large quantities of CO_2, adding to the atmospheric reservoir of the current carbon cycle.

Importantly, that process of burning ancient carbon stocks leaves a distinct fingerprint in the isotopic record as different sources of carbon contain different ratios of the three carbon isotopes—^{14}C, ^{13}C, and ^{12}C (Graven et al., 2020). Over the last 50 years, we can observe a relative decline in heavy ^{13}C relative to light ^{12}C and a rapid decline of ^{14}C, pointing to a source of carbon that comes from terrestrial plants (which are depleted in ^{13}C) and is so old that any ^{14}C it once contained has decayed to non-detectable levels. Only fossil fuels meet all these criteria. Furthermore, when fossil fuels are burned, the carbon bonds with oxygen to form CO_2, which has led to an observable decrease in the proportion of oxygen relative to carbon. Globally, it is clear that the burning of fossil fuels has made up about 87% of CO_2 emissions over the last 20 years (Friedlingstein et al., 2024).

But, there are other sources of GHGs which also need to be addressed. The onset of agriculture over 10,000 years ago began a gradual process of reshaping the Earth's land resources—as humans started cultivating crops, grazing livestock, and cutting down forests for timber. Over the last century, agricultural and urban expansion has grown more intensive, contributing to significant changes in the Earth's land-based carbon sinks, notably forests and peatlands. Today, land use change and deforestation make up around 13% of CO_2 emissions. However, the agricultural sector also produces significant amounts of the other major GHGs, including CH_4 and N_2O, both of which are more potent than CO_2. Whilst their immediate warming potential is greater, offering short-term mitigation potential, they remain in the atmosphere for considerably less time than CO_2, which can hang around for up to 1,000 years.

The result of increasing these GHGs is that the Earth's surface is now 1.47°C warmer than it was in the late 1800s and is hotter than at any other time in the last 100,000 years. But global warming is only the beginning of the story. Because the

Earth is a system where everything is connected, changes in one area can influence changes in all others. Climate change impacts now include intense droughts, water scarcity, severe fires, rising sea levels, flooding, melting polar ice, devastating storms, and declining biodiversity, all of which have potentially catastrophic impacts on human security.

Extreme weather events have always been part of human history, and whilst vulnerable communities have long felt the impact of climate hazards, these events are now being felt more frequently and more widely. A crucial aspect of the climate change debate centres around the fact that despite countries in the Global North, such as the US and European states, contributing the most to historical emissions, it is the most vulnerable peoples of the Global South who are experiencing the greatest impacts. As such, there is a growing recognition of the need for Climate Justice, and to uphold the IPCC concept of 'common but differentiated responsibility,' which recognises that industrialised countries have a greater share of the responsibility, whilst all nations share the impact.

However, the impacts of global warming are not only being felt in the Global South. The European heatwaves of 2022 and 2023, which led to 90,000 people losing their lives (Gallo et al., 2024), have been attributed directly to anthropogenic climate change (Schumacher et al., 2024). They offer a stark reminder to the developed countries that climate change is a global problem. It is no longer a question of mitigation *or* adaptation, but rather an urgent call for transformative change in both rapid decarbonisation and infrastructural and socio-economic adaptation.

Yet, despite the now overwhelming scientific consensus on the drivers, rate and risks of anthropogenic climate change, the global response is still mixed. Decarbonising local, national, and global economies is complex, costly, and rests on dynamic political relationships and commitments. It requires the adoption of new technologies, new ways of allocating capital, new policies and regulations, new ways of doing things, and new cultural norms and expectations. Yet, whilst societies adapt to accelerating climatic and environmental changes, there is a backdrop of increasingly polarised political positions.

In December 2015, at the UN Climate Change Conference (COP21), 196 Parties adopted the Paris Agreement. This landmark treaty offered, for the first time, a binding agreement, bringing all nations together to combat climate change and adapt to its effects. A key element of the agreement was that all parties would submit a Nationally Determined Contribution (NDC), outlining targets and strategies for how each country would cut its carbon emissions. This was a historic, groundbreaking agreement: one that saw virtually every country in the world agree to take substantive action to combat climate change. Another important aspect of the Paris Agreement was that, for the first time, non-Party stakeholders (everyone from businesses to civil society) were invited to start contributing towards climate action.

So, has the Paris Agreement worked? Have governments upheld their NDCs, and has business responded to the call for change? On the one hand, the Paris Agreement put climate change—and the need for countries around the world to tackle it—in the public eye and has led to an increase in efforts to reduce emissions.

But, as of July 2025, only 15 countries met the latest deadline for the submission of NDCs—less than 8% of the 194 parties currently signed up to the Paris Agreement. Without a comprehensive set of NDCs to compare, there will be less pressure on negotiators to raise national ambitions and less political direction from which organisations can take a steer.

Whilst there were some initial signs of positive impact on businesses, fostering low-carbon technologies and markets, and encouraging companies to adopt sustainability goals, the lack of political consensus and the recent rise of climate scepticism as witnessed in Trump's latest withdrawal from the Paris Agreement have stifled climate action. Global spending on the green energy transition is climbing, but the rate of change is slow. The issuance of green bonds in 2024 failed to reach levels seen in 2021, although they are not far off. And whilst emissions continue to grow, with total energy-related emissions increasing by 0.8% in 2024, the International Monetary Fund says that about $5.9 trillion was spent on subsidising the fossil fuel industry in 2020. By 2022, global fossil fuel subsidies had risen to $7 trillion, or 7.1% of GDP, and they are projected to rise again to $8.2 trillion by 2030 (International Monetary Fund, 2023).

The result of such sluggish climate action is that we now face a massive investment gap. At least $4 trillion every year needs to be invested in renewable energy by 2030 (World Economic Forum, 2023). Subsidies need to shift from fossil fuels to renewable energy not only to cut emissions but also to contribute to sustainable economic growth, job creation, better public health, and more equality. The funding is there—what is needed is commitment and accountability, particularly from the global financial systems, including multilateral development banks and other public and private financial institutions.

Having considered the scope and scale of the climate-related problems we are facing, we now turn our attention to the role of organisations and the imperative to become purpose-led.

Organisations in a neoliberal economy and their role in the climate crisis

Organisations represent the coming together of people to undertake tasks and achieve goals that would otherwise not be possible. Whether large or small, commercial, charitable, or publicly funded, organisations are the bedrock of the economy and society. Most of us spend significant amounts of time in and around organisations. At heart, they should be concerned with making life better for everyone—after all, that is their purpose—but many seem to have lost their way. Organisations, particularly corporations within the modern neoliberal economy, with its reliance on burning fossil fuels to create energy as a means of propelling growth, have played a significant role in driving climate change and many societal problems. To understand why this is the case, we need to look at how organisations have evolved in the modern era.

At the risk of oversimplifying a complex trajectory, when corporations first emerged in the Roman era, the state expected them to play a social role and contribute to human well-being. Well-being is defined as fundamental positive outcome states for living beings (PAS 808, 2022). Starting in the 19th century, legislative changes meant that businesses were allowed to incorporate without an explicit obligation to make a positive contribution to society (Davoudi et al., 2018). However, during this era, most firms were either banks or small, family-owned businesses embedded in their local communities and they continued to act pro-socially (Ocasio et al., 2023). Gradually, as industrialisation and technological advances led to an increase in the number and size of corporations through the latter years of the 19th and the early years of the 20th century, a dominant view began to emerge among Western societies that the best way to manage the economy was through a free market with minimal state intervention: the birth of neoliberalism (Mayer, 2021).

Following the Second World War, neoliberalism remained on the fringes of political policy. It was not until the 1970s, after a succession of economic downturns, that neoliberal shareholder capitalism flourished, driven by the views of economists such as Milton Friedman and politicians such as Ronald Reagan in the US and Margaret Thatcher in the UK. Friedman argued that the pursuit of profit was the ultimate goal of the firm, and the role of the state was to get out of the way and allow market competition to decide which should succeed and which should fail. This logic was associated with increasing financialisation of the economy and fostered the view that shareholder value maximisation and growth were the foundation of a flourishing economy.

Thus, the unspoken belief had become embedded by the 1990s that environmental issues should be dealt with through environmental legislation, relations between the employee and the firm should be dealt with via labour law, and competition law should protect and preserve competitive markets (Gartenberg, 2021). In other words, concern for the well-being of people, society, and the planet was no longer viewed as the responsibility of organisations, but rather of governments and legislatures. The assumption was that as the economy and returns to shareholders grew, the benefits would trickle down to everyone. Government planning, state ownership, and collective bargaining were out; unfettered competition, privatisation, and individualism were in. However, appealing as this may sound, in practice, under shareholder capitalism, wealth accrues in the hands of the few, leading to short-termism, a lack of care for people and planet, and rising levels of inequality.

There is no doubt that organisations have been responsible for significant human and social benefits, including improvements in the global standard of living and a reduction in poverty (Foss & Klein, 2024). But the consequences of a neoliberal ideology have also been dire for many people and for the planet we live on; for example, over half of the global population (4.5 billion people) still lack access to essential health services (Mazzucato & Ghebreyesus, 2024). Another example is the ever-widening gap between board member pay and average worker pay: from 1978 to 2023, top CEO pay increased by 1,085% compared with a 24% increase in average workers' pay. In 2023, CEOs were paid 290 times as much as a typical worker, compared with 21 times as much in 1965 (Bivens et al., 2024). Moreover, the richest 1% of people are responsible for 16% of total global emissions, and it would take someone in the bottom 99% 1,500 years to produce as much carbon as the richest billionaires (Oxfam, 2023).

At heart, neoliberalism is premised on the need for constant economic growth. In turn, growth depends on using (or exploiting) nature as a source of raw materials and the means to absorb capitalism's waste (Nyberg & Wright, 2024). Let's take one example: the UK water industry. Back in the 1980s, the UK's state-owned water industry was suffering from years of under-investment with poor-quality drinking water and high levels of pollution on beaches and in waterways. Margaret Thatcher's Conservative government took the decision to privatise the sector and sold water in England and Wales for £7.6 billion as part of a general policy of privatising state industries. The idea was that market forces would drive up quality and help secure the investment needed to meet new EU guidelines. To incentivise the sales, the government took on their historic debt. Company performance would be overseen by a new independent regulatory body, Ofwat. Nearly 40 years later, has this move led to the promised improvements?

While some of the water companies are regarded as well-run, and significant investments have been made in infrastructure, the sector as a whole is in debt to the tune of £60 billion; one company alone, Thames Water, is on the brink of bankruptcy with debts of over £16 billion and has recently received a government bail-out of £3 billion. In 2024, Ofwat announced fines of £168 million for three water firms following serious sewage spills. In 2023, water firms released raw sewage for a total of 3.6 million hours, while some communities have experienced unsafe drinking water or loss of supply. Water leakage rates run at over 3 billion litres per day. Since the companies have an effective monopoly, it has been relatively easy for them to secure loans, which can then be deducted from earnings and reduce their tax bills. The nine major water and sewage companies in England are over 90% owned by overseas investors from countries such as China, Singapore, and Qatar, who therefore have no direct contact with leakages or pollution in the UK.

Moreover, as customer satisfaction levels continue to spiral downwards, between 2021 and 2023, executives at the water companies were collectively paid £70 million, including nearly £41 million in bonuses. According to Ofwat, British water companies have paid out the equivalent of £78 billion in dividends to shareholders since 1990. Ofwat itself has been criticised for failing to ensure sufficient investment was being made to ensure the infrastructure could cope with changes in the weather, such as increased periods of heavy and intense rain, and to ensure the long-term sustainability of the water firms. Water UK, the trade association for the water industry, has said that the 'current system is not working and needs major reform'; at the time of writing, a government commission is undertaking a wholesale review of the sector (CIWEM, 2019; Fisher, 2025; Hansard, 2024; Jack, 2024). The case of the water industry exemplifies shareholder capitalism's preference for short-term profit at the expense of long-term well-being.

Shareholder capitalism also needs growth, and growth needs energy—burning fossil fuels is a tried-and-tested, familiar, and comparatively cheap way of generating that energy. So reliant are we on fossil fuels that the sector is the most heavily capitalised segment of the economy worldwide, and the main focus of global investment on the part of banks, insurance companies, and pension funds (Nyberg & Wright, 2024). Although there has been strong scientific consensus on fossil fuels driving climate change since the 1990s (Yale, 2025), organisations in the fossil fuel sector have deliberately sought to confuse the issue through spreading misleading data and undertaking intensive political lobbying. In 2022, oil and gas companies in the US reportedly spent $124.4 million lobbying central government (Sayki & Cloutier, 2023). It is also estimated that the 2024 UN climate summit COP29 was attended by over 1,750 coal, oil, and gas lobbyists, outnumbering the delegations sent by almost every country (Noor, 2024).

Many other social and health-related problems have also been laid at the door of profit-making organisations and their lobbying, including the obesity and opioids crises (Davis, 2021). As Davis (2021, p. 904) writes, 'nearly every major societal pathology in the West today . . . is caused or exacerbated by profit-driven corporations.' It is no surprise therefore that confidence in business is at an all-time low: one study showed just one-third of adults believe business is good for society (O'Brien, 2023).

Although the millions spent lobbying might buy political influence—who can forget US President Donald Trump's exhortation to 'drill, baby, drill!'—the oil companies have had much less success in persuading citizens. A recent survey of 130,000 people in 125 countries across the world found that 89% want governments to do more to combat global warming, with over 2/3 saying they are willing to give up 1% of their income to help with this endeavour (Carrington, 2025). Even supporters of political parties that do not typically favour climate mitigation policies are supportive—the study found that 62% of right-wing Reform UK voters said they thought it was important that government cares about tackling climate change. Research has shown widespread support for action to address climate change, so long as these maintain freedom of choice, meet the conditions of fairness and accessibility, and provide additional benefits such as health, cost of living, local environment, or waste reduction (Cherry et al., 2024). This 'silent majority' deserves to know that their views are widely shared by others, and to be listened to by governments and organisations.

While firms lobby and jostle to keep the pipelines open, it is ironic to note that the projected impacts of climate change on organisations themselves are likely to be devastating. The World Economic Forum (2024) estimates that emissions already in the atmosphere will lower gross domestic product (GDP) per capita by between 11% and 29% by 2050. Climate hazards could drive between $560 and $610 billion of fixed asset losses by 2035, representing a drop of between 6.6% and 7.3% in average company earnings, with heatwaves alone estimated to depress productivity by $7.1 trillion by 2050. In situations where multiple businesses are hit simultaneously by a climate-related event, the impacts can cascade through whole industries and regions, leading to direct operational costs, supply chain disruptions, infrastructure damage, physical health impacts, and societal instability. As the World Economic Forum states (2017, p. 1), 'neoliberal economics has reached breaking point.' So, how can we do things differently and, specifically, what can organisations do?

Organisational purpose: What is it and how can it help?

It is against this backdrop that purpose as an alternative logic has come to the fore in recent years, with increasingly vociferous calls for organisations to step up and take their share of responsibility for the long-term well-being of people and the planet by becoming purpose-led. These calls have come not just from climate activists but from political parties and governments of all persuasions, regulators, investors, royalty, academics, institutions such as the World Economic Forum and the United Nations, media outlets including the Economist and the Financial Times, as well as employees, customers, and communities, and organisational leaders themselves. Many organisations are now responding to this call by developing a statement of purpose, recognising that humans are the environmental stewards of the planet and have an obligation to protect it for future generations (Hoffman, 2017).

Two public pronouncements are often cited as emblematic of this change in direction. In 2018, Larry Fink, CEO of investment firm BlackRock, famously penned a letter to leading CEOs calling on companies to 'serve a social purpose,' while in 2019, the Business Roundtable of 181 prominent CEOs in the USA issued a statement committing to lead their companies for the benefit of all stakeholders, a reversal of their 1997 statement which proclaimed the purpose of the organisation to be profit maximisation (Business Roundtable, 2019). In the UK, the British Academy funded a major research programme examining how business could be reconceptualised around purpose (British Academy, 2019).

But what exactly is an organisational purpose? Purpose has been defined in several ways, and this has been the cause of some confusion. George and colleagues (2023) helpfully distinguish between two types of purpose:

- *Goal-based purpose*. This basic view of purpose suggests that purpose simply means an overarching strategic goal or aim pursued by an organisation, such as brewing the best local craft ale or providing broadband services at a competitive price. However, this can encompass almost any kind of purpose, not necessarily one that is going to address any of the fundamental problems we face today, and as a result, it does not represent any challenge to the status quo. This type of purpose is not the focus of this book.
- *Duty-based purpose*. A duty-based purpose is linked to the organisation's moral obligation to support long-term well-being for people and the planet. Some organisations were established specifically with this in mind. A well-known example is outdoor clothing manufacturer Patagonia, which from the outset aimed to use materials that caused least harm to the environment in their production processes. In 2018, the company proclaimed its purpose as 'we're in business to save our home planet.' Founder Yvon Chouinard recently transferred 100% of the company's voting stock to a Trust established specifically to protect the company's values with Earth 'now our only shareholder' (Patagonia, 2022). Most organisations, however, were not founded with purpose in mind, and so leaders are having to work out what changes are needed in order to become purpose-driven.

Although there are many definitions, the British Standards Institute's (BSI) definition of purpose is one that is being widely adopted. According to the BSI, purpose-driven organisations are defined as organisations that exist

> to achieve an optimal strategic contribution to the long-term wellbeing of all people and planet . . .[and] make sure that, at a minimum, and regardless of its purpose, they make a fair and adequate contribution to ensuring that the systems, capitals and stakeholders they rely on are healthy.

They should also 'approach these two tasks with decision-making that is both wise . . . and ethical' (PAS 808, 2022, p. v) and therefore 'not profit from producing problems for people or planet' (Mayer, 2021). Becoming purpose-driven therefore entails establishing a core well-being purpose that determines the strategic direction of the organisation. At the same time, this purpose should ensure the organisation eradicates any actions or behaviours that could cause harm to either people or the planet. Profit then becomes secondary to purpose and not the driver of organisational actions and decision-making. This reorientation represents a radical departure from many of the current assumptions about how the economy works, and the role of organisations within that (Hurth & Stewart, 2022). It reminds us that organisations exist with society's permission and in service to society, and are not simply vehicles to make a profit and serve the short-term interests of owners and managers.

It is important to note at this point that an organisational purpose can relate to other forms of well-being than planetary health—the enhancement of human health, equality, well-being, or education, for example. In 2015, the United Nations developed 17 Sustainable Development Goals aimed at eliminating poverty, preserving the planet, and ensuring prosperity; and purpose can be oriented towards any of these. The chapters relating to transforming organisations to become purpose-led in this book apply just as well to these types of purpose as they do to purposes linked to climate change. We are focusing our discussion of purpose within the context of climate change, however, because of the fundamental threat that it poses to the sustainability of life on Earth. Even where an organisation chooses to focus on other core purposes, say, education, it still has a responsibility to ensure that its actions at least cause no harm to the planet.

Purpose is thus more than an extension of current thinking; it is a complete rethink about how organisations could and should work and how they can contribute to human well-being (Mayer, 2021). Under purpose-led approaches, crucially, profit becomes a by-product of purpose, not the main aim. Purpose also represents an advance over other approaches to considering the organisation's responsibilities to society and the planet, such as stakeholder capitalism or corporate social responsibility (CSR). Stakeholder capitalism argues that organisations should take into account the needs and wishes of multiple stakeholders in and around the organisation when making decisions, such as employees, customers, and communities (Ocasio et al., 2023).

However, it offers no solution to the difficulties of managing trade-offs between various stakeholder groups and does not necessarily consider the natural environment and planetary boundaries (Bansal et al., 2024). CSR programmes can often fall into the trap of becoming a series of ad hoc, peripheral actions that may assuage any misgivings about the harms caused by corporate actions, but largely do not challenge the status quo in terms of shareholder value maximisation (PAS 808, 2022). Both these viewpoints continue to place the organisation and profit-making front and centre of decision-making, whereas purpose decentres the organisation and considers it as just one entity within a wider ecosystem.

Sceptics will be reassured by emerging research findings which suggest that authentically purpose-driven organisations enjoy a number of benefits, including more positive consumer evaluations and increased ability to attract talented workers, as well as higher levels of performance and innovation (George et al., 2023; Ibn-Mohammed et al., 2024). However, we cannot pretend that transforming an organisation to become purpose-driven is easy or straightforward; whatever you may have read to the contrary, there is no quick, ten-step solution that everyone can follow. To understand why this is the case, we next turn our attention to the challenges organisations are likely to face on the way to becoming purpose-driven.

Challenges on the way to becoming purpose-driven

Although we, the editors, along with the other contributors to this book, argue strongly that organisations need to change, we are all keenly aware that not everyone agrees with us, and equally that successfully turning an organisation around is no easy undertaking. The shift can entail high development costs, uncertain outcomes, and potentially lower levels of profitability in the short term (Schuper & Soppe, 2025), as well as opposition from powerful stakeholders. There have been some recent high-profile instances of companies well-known for their environmental credentials that have then drastically scaled down or withdrawn their plans, such as Unilever, which can cause others to follow suit.

Challenges can arise at a number of levels, and we briefly summarise some of these as follows.

Macro-level challenges

When BP announced in 2023 that it was reversing its decarbonisation goals, the company's share price rose by 7% on the day of the announcement (Davis & DeWitt, 2024). When Unilever's CEO announced he would discontinue quarterly earnings reports to facilitate their Sustainable Living Plan, the company lost over $2 billion in market capitalisation (Besharov & Mitzinneck, 2023). The pressure exerted by business-as-usual (BAU) thinking on organisations is immense. As Davis (2021, p. 907) argues, 'when purpose and shareholder value get into a boxing ring, I will bet on shareholder value every time.' He goes on to note that the operation of the neoliberal economy and its various institutions combine to reward and encourage corporations that provide strong shareholder returns over and above anything else. With the best of intentions, companies find it hard to resist the call back to growth and profit maximisation.

Even the way we talk about climate change and organisational purpose influences the way people feel about them at an individual level and determines what is seen as 'normal' at a societal level. Academics with an interest in 'discourse analysis,' or the study of how talk, belief, and actions are linked, have found that 'discourses of delay' relating to climate change are widespread in the media and in political debates (Cherry et al., 2024). While climate change denial has significantly decreased, these discourses of delay can foster overconfidence in current actions, redirect responsibility to 'other people,' or lead to dejection—and inaction—in the face of the scale of the changes required. All of these can impede progress towards a solution.

This does not mean that organisations cannot become purpose-driven due to factors beyond their control; organisations are a core part of the system and can act to exert change not just internally but also systemically. It is worth remembering the significant shifts that have taken place in the corporate world since the 20th century (Mayer, 2021). While small to medium-sized companies (SMEs) account for 99% of all businesses in Europe and employ around 100 million people (Ibn-Mohammed et al., 2024), some modern corporations are vast. Companies such as Apple, Microsoft, and Amazon have a market capitalisation equivalent to the GDP of some major countries; by 2020, Apple had reached a capitalisation of US$2 trillion, surpassing the GDP of countries such as Italy, Brazil, Canada, and Russia (Patriotta, 2021). Firms such as these are increasingly complex and increasingly global, and have the power to promote a shift in the entire system—if they choose to do so. Organisations of all sizes can control what they prioritise, how they allocate resources, how they reward executives and employees, and how they contribute to society and the planet.

Organisational challenges

It is one thing for leaders to decide they want their organisation to become purpose-driven, but it is quite another for this to be turned into a reality. Developing a statement of purpose can itself be an exercise fraught with political infighting and disagreements. The resulting statement may miss the mark—a recent study by Christopher Michaelson and colleagues (2020) of 2000 purpose statements found that 93% lacked any meaningful sense of purpose. There is also the risk of purpose-washing, or making empty claims about purpose (AlSalim & Etter, 2023)—disappointingly, some of the signatories of the 2019 Business Roundtable statement have themselves been criticised for excess carbon emissions and sponsoring fossil fuel campaigns (Kaplan, 2023). Other challenges include securing buy-in for purpose, managing the transformation to become purpose-driven, communicating the purpose, avoiding short-term pressures, and finding appropriate metrics (Giorgi, 2017; Ibn-Mohammed et al., 2024).

However, while organisations hesitate, employee activism around purpose is on the increase. More and more employees are grouping together and demanding that their employers live up to their values. Some recent examples include:

- Facebook employees staged a virtual walkout due to the company's tolerance of racism and hate speech
- 4,000 Amazon employees filed a resolution criticising their company's stance on climate change
- Staff at US IT firm Salesforce wrote to their CEO asking him to sever company ties with Customs and Border Protection, the agency responsible for detaining young children on the Mexico border
- 20,000 Google employees walked out in protest about the company's sexual harassment policies (Briscoe & Gupta, 2021; Davis, 2021)

Employee pressure is both a challenge and an opportunity for organisations; having a strong and authentic purpose is one way of attracting and retaining talented staff. A recent survey found that seven out of ten employees of all ages in the UK were worried about climate change, with one-third of workers willing to quit if they disagreed with their employer's stance on climate change and 48% saying they would never consider a job in certain sectors for environmental or moral reasons, whatever the pay (Rowsell, 2023).

Guide to the book

Who is this book for?

The book is for anyone interested in how organisations can become purpose-led. In particular, we have in mind purpose professionals and other senior leaders seeking to drive forward purpose in their organisation, whether in the public, private, or not-for-profit sectors. It is equally relevant to leaders of small to medium-sized organisations (SMEs). Students at undergraduate and postgraduate levels studying modules in purpose, corporate social responsibility, or climate change from an organisational perspective will also find much that is relevant to them, as will leaders attending executive education programmes focusing on these topics. Policy-makers may also find the book helps them understand the challenges organisations face in addressing climate change.

It was important to us that the contributors to the book were leading experts in their field, and that the book included multidisciplinary perspectives, reflecting the complexity of the topic—but written in such a way that the non-expert can understand. This has led to a unique collaborative effort that has not been without its challenges. Both editors worked with the authors of all the chapters, and we soon found that even simple words, such as 'organisation,' could mean different things to people approaching the topic from different perspectives. However, the result is a book that brings together in one place the latest thinking and evidence-based recommendations that can help organisations address the most complex challenges of today.

Summary of the book

In Chapters 1–3, we set the context and explain the evidence as well as the causes and consequences of climate change in non-scientific terms—useful for people looking for evidence to help persuade the sceptics they work with. Chapter 4 explores the shift from shareholder to stakeholder capitalism, and then in the remaining chapters, we focus on organisational purpose itself—we explain what it is, why it matters, and what the research evidence tells us about how to embed it across the organisation. These latter chapters are relevant regardless of whether your purpose focuses on the climate crisis or other areas of the Sustainable Development Goals.

Each of the topics covered is significant in its own right, and some areas are beyond the scope of what we can cover. For example, we only touch on the broader economic context within which organisations are operating and the economic transformation options that are available, such as degrowth or sustainable growth, although we know these to be extremely important issues that directly affect organisations. We also do not go into detail on what, specifically, an organisation's purpose should be, although we do talk about the process of developing a purpose (in Chapter 5), in large part because each organisation will need to develop its own particular purpose that is relevant to its history, sector, and context. We also only briefly discuss the topic of corporate governance, although, again, corporate governance models will need to be reconsidered as we move towards purpose-driven organisational models.

Finally, it is important to note that this is not intended to be a textbook, written in a consistent way across each chapter. The chapters have a similar structure in that they include a summary of the key points of their chapter, examples, case studies, and suggestions for further reading, but each chapter is written by a different author team chosen for their expertise, each with their own distinctive voice.

Chapter summaries

Chapter 1 sets the foundations of evidence for contemporary climate change. Drawing on material from his popular science books *How to Save Our Planet: The Facts,* and *Climate Change, A Very Short Introduction,* Mark Maslin describes the Earth's natural greenhouse effect, explaining, in simple terms, the way in which greenhouse gases, such as CO_2 and CH_4, warm our atmosphere. Maslin describes how climates have changed in the past and the striking co-variance between GHGs and temperature, and how we have been able to separate natural cycles of climate change from human-induced—anthropogenic—global warming. Maslin provides critical insight into the history of climate science and why there has been a delay in accepting the evidence. He concludes the chapter with some stark reminders of what our future may look like under different warming scenarios.

In Chapter 2, Anita Wreford and Katie Manning consider the socio-economic drivers of climate change and the impacts it is having on communities at different scales. They consider the effects of global warming on biodiversity loss and the broader impacts on agricultural productivity and food security. Key to understanding the effects of climate change is recognising that changes in weather systems and impacts on the primary sector, notably on food production and water, are interconnected across the human system. This leads to increased risk for human security, displacement, labour markets, and infrastructure. Such knock-on effects pose a significant threat to business operations and global GDP. How we respond to the climate crisis will also have its own socio-economic impacts, and considering the wider systemic trade-offs in any climate solution is critical for sustainable development. Throughout this chapter, Wreford and Manning outline some specific organisational actions for mitigating specific climate change impacts. The chapter concludes with a dedicated section on Nature-based solutions, and the implementation of cost-effective nature-positive action.

In Chapter 3, Duncan Russel and Nick Kirsop-Taylor explore the global, national, and local policy-oriented organisations of climate governance. They highlight the way in which climate action is underpinned by policies that operate at different scales of governance. As the umbrella organisation for climate action, the United Nations Framework Convention on Climate Change (UNFCCC) coordinates cross-border political efforts to combat global warming. Yet decision-making by state actors remains somewhat limited by a lack of coordination at the national scalc. Russel and Kirsop-Taylor demonstrate the challenges facing international cooperation, highlighting the detrimental impacts of economic competitiveness concerns and disputes over historical responsibilities for carbon emissions. In recent years, however, there has been an explosion of organisations that actively engage with the UNFCCC, including non-governmental organizations (NGOs), businesses, investors, and other non-state actors, who work to align their activities with international targets. Critically, Russel and Kirsop-Taylor draw attention to the increasing complexity of the policy-oriented organisational ecosystem, and both the risks and opportunities that such complexity creates.

In Chapter 4, Rashedur Chowdhury considers the difference between shareholder capitalism and stakeholder capitalism, setting the scene for the subsequent chapters that delve more deeply into purpose-led transformation. Shareholder capitalism has dominated economic thinking for decades. It is based on the premise that the main function of organisations is to make a profit and generate returns for shareholders. An alternative view is stakeholder capitalism, under which organisations are required to consider the interests of a wide range of stakeholders, including employees, customers, suppliers, and communities, in their decision-making. In the chapter, Chowdhury argues that stakeholder capitalism does not go far enough, because it still fails to position stakeholders, especially marginalised and vulnerable stakeholders such as indigenous communities and the environment, front and centre of the organisation's concerns. Chowdhury explains the implications of moving beyond a stakeholder perspective and outlines how organisations can work with vulnerable groups to ensure they place the needs of marginalised communities and the environment at the heart of their decision-making.

Victoria Hurth and Charles Ebert have been deeply involved in the development of the British Standard in Purpose-Driven Organisations and have many years of experience in research, policymaking, and organisational consultancy relating to purpose. They draw on these experiences in Chapter 5, where they set out a definition of organisational purpose and explain the history of the development of the concept. They explain why becoming purpose-led represents a paradigm shift away from 'business as usual' and short-term financial self-interest towards a new modus operandi for organisations that places collective long-term well-being as the core purpose and aim for all organisations. In doing so, they situate the emphasis on purpose within the broader context of the multiple and interconnected threats to planetary, environmental, and human sustainability we are facing today. Hurth and Ebert also provide guidance for organisations in developing their purpose statement, and explain the interlinkages between purpose, strategy, values, and mission.

In Chapter 6, Craig V. Robinson and Luis D. Torres consider how organisations can manage the change process that becoming purpose-led entails. Recognising that the process is likely to be messy and challenging at times, and may well involve dealing with different forms of resistance, they put forward six recommendations for achieving a successful transformation. These include identifying your starting point, considering how to approach resistance, empowering leaders at all levels, engaging key stakeholders, embedding purpose through the organisational culture, and measuring and reporting on performance. They emphasise the importance of transparent communication, aligning incentives and empowering change agents through the organisation, as well as reinforcing accountability for purpose implementation.

Catherine Tilley has conducted research into how purpose can be integrated into decision-making and behaviours at all levels in the organisation. In Chapter 7, she emphasises the importance of ensuring that the formal and social structures of the organisation both work to ensure that purpose is front and centre of people's minds when making decisions. The default for most organisations is to base decisions on purely financial considerations; the shift to becoming purpose-led therefore represents a significant change that needs to be embedded within all decision-making pathways. One particular challenge is that organisational purposes can be complex, with tensions potentially arising between different stakeholders over what is the 'right' decision to make. Organisations can be held back by the availability of appropriate data, conventional decision-making methods that are no longer fit for purpose, and the habits of mind of decision-makers. Tilley advocates a holistic approach that takes account of the organisation's external as well as internal context, and the characteristics of decision-makers themselves, such as unconscious and conscious biases. She puts forward a number of solutions to these issues and shows how organisational systems can be adapted to place purpose at the forefront of people's minds in the decision-making process.

Chapter 8, written by human resource management expert Na Fu, focuses on the HRM implications of the transformation to become purpose-led. Fu explains how purpose needs to be integrated into the HRM life cycle from recruitment and onboarding, through learning and development, performance management, rewards, involvement, and exit. She outlines five core challenges that the HR function faces in embedding purpose into their operations: balancing purpose and profit; integrating consistent approaches to all employees with the need to personalise their offering; leveraging technologies and analytics; ensuring the ethical and legal use of technologies in supporting people management; and fostering cross-functional leadership collaboration to ensure that purpose becomes embedded across the organisation.

While having a purpose statement is fundamental to developing the purpose-driven organisation, as we saw in Chapter 5, unless there is effective communication around the purpose, then it risks becoming an empty slogan. In Chapter 9, organisation communication experts Joep Cornelissen and Mirjam Werner examine purpose communication at all stages from the initial development of the purpose statement through to embedding the purpose over time. In particular, they explain the key role that a purpose narrative can play in bringing the purpose to life. A purpose narrative comprises three elements: the aspiration story, the character story, and the stakeholder story. As such, the narrative serves to make the purpose meaningful to key stakeholders. Cornelissen and Werner emphasise the importance of constant consultation and dialogue around the purpose with both employees and external stakeholders to foster transparency and ensure the authenticity of the purpose.

Finally, in Chapter 10, David Aikman and Marc Lepere consider how organisations can go about reporting on the progress they are making to become purpose-led. They position their chapter in the context both of the growing risks posed to organisations by the climate crisis, and of the developing backlash against climate action, which is leading to a breakdown of the global consensus and confusion over how to account for, measure, and report on purpose. They show how firms are responding in three different ways to the situation, with some focusing on compliance, others integrating climate risk into decision-making, and a final group turning to purpose as an opportunity to attract capital and enhance long-term value. In the chapter, Aikman and Lepere argue that the voluntary reporting standards relating to climate risks that have dominated thus far are now being questioned. Efforts to agree on more encompassing mandatory reporting standards have yet to lead to a clear international consensus. Meanwhile, there are emerging calls around the world to develop specific standards relating to organisational purpose, with an emphasis on regeneration and long-term stakeholder value, drawing on the UK's PAS808: 2022 voluntary standards as an exemplar. The challenge is to develop clear and consistent metrics that enable social and environmental outcomes to be captured with accuracy. Aikman and Lepere also consider the growing risk that climate-related litigation poses for firms that fail to abide by legal requirements, exaggerate their environmental credentials, or neglect to take environmental impacts into consideration. Drawing on a decade of experience in climate finance, the authors outline eight decision-critical questions that business leaders need to consider.

Conclusion

In this book, we hope to inform organisations about how the climate crisis is affecting the economy, and to inspire them with insights into how to transform to become purpose-driven. Changing the way our organisations work, their priorities, and their role in society is not easy, but it is possible. As Ibn-Mohammed and colleagues (2024) write, we need courageous leaders to step up, challenge the status quo, and lead us towards a brighter future for humanity and for the planet.

References

AlSalim, F., & Etter, M. (2023). *Purpose-washing: What it is and how to avoid it*. King's Business School Meaning and Purpose Network.

Bansal, P., Durand, R., Kreutzer, M., Kunisch, S., & McGahan, A. M. (2024). Strategy can no longer ignore planetary boundaries: A call for tackling strategy's ecological fallacy. *Journal of Management Studies*. https://https://doi.org/10.1111/joms.13088

Besharov, M., & Mitzinneck, B. (2023). The multiple facets of corporate purpose: An analytical typology. *Strategy Science*, *8*(2), 121–321. https://doi.org/10.1287/stsc.2023.0186

Bivens, J., Gould, E., & Kandra, J. (2024). *CEO pay declined in 2023 but it has soared 1,085% since 1978 compared with a 24% rise in typical workers' pay*. Economic Policy Institute. https://www.epi.org/publication/ceo-pay-in-2023/

Briscoe, F., & Gupta, A. (2021, Winter). Business disruption from the inside out. *Stanford Social Innovation Review*. https://ssir.org/articles/entry/business_disruption_from_the_inside_out

British Academy. (2019). *Principles for purposeful business: How to deliver the framework for the future of the corporation*. British Academy.

Business Roundtable. (2019, August 19). Business Roundtable redefines the purpose of a corporation to promote 'an economy that serves all Americans'. *Business Roundtable*. https://www.businessroundtable.org/business-roundtable-redefines-the-purpose-of-a-corporation-to-promote-an-economy-that-serves-all-americans

Carrington, D. (2025, April 22). 'Spiral of silence': Climate action is very popular, so why don't people realise it? *Guardian*. https://www.theguardian.com/environment/2025/apr/22/spiral-of-silence-climate-action-very-popular-why-dont-people-realise

Cherry, C., Verfuerth, C., & Demski, C. (2024). Discourses of climate inaction undermine public support for 1.5°C lifestyles. *Global Environmental Change*, *87*, 102875. https://doi.org/10.1016/j.gloenvcha.2024.102875

CIWEM. (2019). *Thirty years on, what has water privatisation achieved?* https://www.ciwem.org/the-environment/how-should-water-and-environmental-management-firms-tap,-retain-and-promote-female-talent

Cloutier, C., Desjardins, F., & Rouleau, L. (2024). Grand challenges viewed through the pragmatist lens of the economies of worth: A multidisciplinary review and framework for the conduct of moral work in pluralistic settings. *Journal of Management Studies*. https://doi.org/10.1111/joms.13078

Concern UK. (2024, January 2). *I have experienced the impact of climate change first hand. This is the reality*. https://www.concern.org.uk/news/climate-change-kenya-experience

Davis, G. F. (2021). Corporate purpose needs democracy. *Journal of Management Studies*, *58*(3), 902–913. https://doi.org/10.1111/joms.12659

Davis, G. F., & DeWitt, T. (2024). Can strategy address the climate crisis without losing its essence? *Journal of Management Studies*. https://doi.org/10.1111/joms.13083

Davoudi, L., McKenna, C., & Olegario, R. (2018). The historical role of the corporation in society. *Journal of the British Academy*, *6*(1), 17–47. https://doi.org/10.5871/jba/006s1.017

Dickinson, D. (2024, August 26). What is sea level rise and why does it matter to our future? *United Nations World*. https://news.un.org/en/story/2024/08/1153596

Fink, L. (2018, January 17). A sense of purpose. *Harvard Law School Forum on Corporate Governance*. https://corpgov.law.harvard.edu/2018/01/17/a-sense-of-purpose/

Fisher, D. (2025, February 10). What public-private-partnership scandals can tell us about wrongdoings in the water industry. *The Conversation*. https://theconversation.com/what-public-private-partnership-scandals-can-tell-us-about-wrongdoing-in-the-water-industry-249218

Fitzgerald, J. (2025, January 9). What's the latest on Los Angeles wildfires and how did they start? *BBC News*. https://www.bbc.co.uk/news/articles/clyxypryrnko

Foss, N. J., & Klein, P. G. (2024). Do we need a 'new strategy paradigm'? No. *Journal of Management Studies*. https://doi.org/10.1111/joms.13081

Friedlingstein, P., O'Sullivan, M., Jones, M. W., Andrew, R. M., Hauck, J., Landschützer, P., . . . & Zeng, J. (2024). Global carbon budget 2024. *Earth System Science Data Discussions*, 1–133. https://doi.org/10.5194/essd-17-965-2025 (Original work published 2025.)

Gallo, E., Quijal-Zamorano, M., Méndez Turrubiates, R. F., Tonne, C., Basagaña, X., Achebak, H., & Ballester, J,. (2024). Heat-related mortality in Europe during 2023 and the role of adaptation in protecting health. *Nature Medicine*, *30*, 3101–3105. https://doi.org/10.1038/s41591-024-03186-1

Gartenberg, C. (2021). Purpose-driven companies and sustainability. In *Handbook on the business of sustainability* (Chapter 2, pp. 24–42). Edward Elgar.

George, G., Haas, M. R., McGahan, A. M., Schillebeeckx, S. J. D., & Tracey, P. (2023). Purpose in the for-profit firm: A review and framework for management research. *Journal of Management*, *49*(6), 1841–1869. https://10.1177/01492063211006450

Giorgi, S. (2017). The mind and heart of resonance: The role of cognition and emotions in frame effectiveness. *Journal of Management Studies*, *54*(5), 711–738. https://doi.org/10.1111/joms.12278

Graven, H., Keeling, R. F., & Rogelj, J. (2020). Changes to carbon isotopes in atmospheric CO2 over the industrial era and into the future. *Global Biogeochemical Cycles*, *34*(11), e2019GB006170. https://doi.org/10.1029/2019GB006170

Hansard. (2024, February 22). *Water and sewage companies: Directors' remuneration* (Vol. 836). https://hansard.parliament.uk/lords/2024-02-22/debates/55D8B31D-41B2-4991-80D7-1F61127249BA/WaterAndSewageCompaniesDirectors

Hoffman, A. J. (2017). *Finding purpose: Environmental stewardship as a personal calling*. Routledge.

Hurth, V., & Stewart, I. S. (2022, January 5). Re-purposing universities: The path to purpose. *Frontiers in Sustainability*. https://doi.org/10.3389/frsus.2021.762271

Ibn-Mohammed, T., Herath, S., Swallow, N., Gower, C., Montes, A. P., Brooks, D., Garcia, A. P., & Morrison-Saunders, A. (2024). Overcoming the challenges faced by purpose-led medium-sized businesses: The role of corporate communication and its interplay with corporate reputation and organisational identity. *Corporate Reputation Review*. https://doi.org/10.1057/s41299-024-00181-7

International Monetary Fund. (2023). *Fossil fuel subsidies*. https://www.imf.org/en/Topics/climate-change/energy-subsidies

Jack, S. (2024, October 25). The water industry is in crisis. Can it be fixed? *BBC News*. https://www.bbc.co.uk/news/articles/c0qdev4vyl5o#:~:text=Millions%20of%20customers%2C%20surfers%20and,not%20hard%20to%20see%20why.&text=The%20independent%20commission%20will%20be,for%20customers%20and%20their%20bills?

Kaplan, S. (2023). The promises and perils of corporate purpose. *Strategy Science*. https://doi.org/10.1287/stsc.2023.0187

Lyons, K. (2022, November 8). How to move a country: Fiji's radical plan to escape rising sea levels. *Guardian*. https://www.theguardian.com/environment/2022/nov/08/how-to-move-a-country-fiji-radical-plan-escape-rising-seas-climate-crisis

Mayer, C. (2021). The future of the corporation and the economics of purpose. *Journal of Management Studies*, *58*(3), 887–901. https://doi.org/10.1111/joms.12660

Mazzucato, M., & Ghebreyesus, T. A. (2024). Advancing the economics of health for all. *The Lancet, 404,* 998–1000. https://doi.org/10.1016/S0140-6736(24)01873-7

Michaelson, M., Lepisto, D. A., & Pratt, M. G. (2020, August 17). Why corporate purpose statements often miss their mark. *Strategy + Business*. https://www.strategy-business.com/article/Why-corporate-purpose-statements-often-miss-their-mark

Noor, D. (2024, November 15). Over 1,700 coal, oil and gas lobbyists granted access to COP29, says report. *Guardian*. https://www.theguardian.com/environment/2024/nov/15/coal-oil-and-gas-lobbyists-granted-access-to-cop29-says-report

Nyberg, D., & Wright, C. (2024). Confronting the climate crisis: Fossil fuel hegemony and the need for decarbonization, degrowth, and democracy. *Journal of Management Studies*. https://doi.org/10.1111/joms.13198

O'Brien, A. (2023, November). *The purpose dividend*. Demos B Lab UK.

Ocasio, W., Kraatz, M., & Chandler, D. (2023). Making sense of corporate purpose. *Strategy Science*. https://doi.org/10.1287/stsc.2023.0054

Oxfam. (2023). *Richest 1% emit as much planet-heating pollution as two-thirds of humanity*. https://www.oxfam.org.uk/mc/qer7km/

PAS 808. (2022). *Purpose-driven organizations—worldviews, principles and behaviours for delivering sustainability—guide*. British Standards Institute.

Patagonia. (2022). *Ownership*. https://eu.patagonia.com/gb/en/ownership/

Patriotta, G. (2021). The future of the corporation. *Journal of Management Studies*, *58*(3), 879–886. https://doi.org/10.1111/joms.12673

Rowsell, J. (2023, September 25). Two in five over 55s would climate snub a firm with the wrong ESG credentials, survey reveals. *People Management*. https://www.peoplemanagement.co.uk/article/1838501/two-five-55s-climate-snub-firm-wrong-esg-credentials-survey-reveals

Sayki, I., & Cloutier, J. (2023, February 22). Oil and gas industry spent $124.4 million on federal lobbying amid record profits in 2022. *Open Secrets*. https://www.opensecrets.org/news/2023/02/oil-and-gas-industry-spent-124–4-million-on-federal-lobbying-amid-record-profits-in-2022/

Schumacher, D. L., Zachariah, M., Otto, F., Barnes, C., Philip, S., Kew, S., Vahlberg, M., Singh, R., Heinrich, D., Arrighi, J., van Aalst, M., Hauser, M., Hirschi, M., Bessenbacher, V., Gudmundsson, L., Beaudoing, H. K., Rodell, M., Li, S., Yang, W., . . . Seneviratne, S. I. (2024). Detecting the human fingerprint in the summer 2022 western–central European soil drought. *Earth System Dynamics*, *15*(1), 131–154. https://doi.org/10.5194/esd-15-131-2024

Schupfer, H., & Soppe, B. (2024). Greening from within: The role of organisational purpose shift in building internal legitimacy for fossil fuel incumbents' green innovation. *Industry and Innovation*. https://doi.org/10.1080/13662716.2024.2379995

Sherriff, L. (2025, January 10). 'I have nothing to go back to'—LA fires heartbreak. *BBC News*. https://www.bbc.co.uk/news/articles/c0eweel5zg3o

Statista. (2025). *Annual carbon dioxide (CO2) emissions worldwide from 1940 to 2024*. https://www.statista.com/statistics/276629/global-co2-emissions/

UKGovernment. (2022). *UK climate risk assessment 2022*. https://assets.publishing.service.gov.uk/media/61e54d8f8fa8f505985ef3c7/climate-change-risk-assessment–2022.pdf

World Economic Forum. (2017, July 17). *The end of neoliberalism?* https://www.weforum.org/stories/2017/07/this-is-what-the-future-of-economic-liberalism-looks-like-its-time-to-rethink-it/

World Economic Forum. (2023). *IEA: Clean energy investment must reach $4.5 trillion per year by 2030 to limit warming to 1.5°C*. https://www.weforum.org/stories/2023/09/iea-clean-energy-investment-global-warming/

World Economic Forum. (2024, December 11). *Business on the edge: Building industry resilience to climate hazards*. https://www.weforum.org/publications/business-on-the-edge-building-industry-resilience-to-climate-hazards/

World Meteorological Association. (2025, January 10). *Wildfires rage in California*. https://wmo.int/media/news/wildfires-rage-california

Wilson, O. J., & Pescott, O. L. (2023). Assessing the exposure of UK habitats to 20th- and 21st-century climate change, and its representation in ecological monitoring schemes. *Journal of Applied Ecology*, *60*, 1995–2006. https://doi.org/10.1111/1365-2664.14455

Yale. (2025, April 25). *Scientists knew about global warming in the 1950s*. https://yaleclimateconnections.org/2025/04/scientists-knew-about-global-warming-in-the-1950s/

1 The science of human-induced climate change

Mark Maslin

Key points

- Carbon dioxide has risen by over 50% and methane by 150% since pre-industrial times, raising global temperatures by almost 1.5°C. The main drivers of recent global warming are fossil fuel use and deforestation
- Despite climate change being an old science, global warming theory only came to prominence in the 1980s
- The Intergovernmental Panel for Climate Change (IPCC) was established in 1988 and, since then, has published six comprehensive Assessment Reports and 15 Special Reports, detailing the current knowledge on climate change, its causes, potential impacts, and response options
- If global warming stays consistently above 1.5°C, it is likely that several climate tipping points will pass their critical threshold, including the collapse of the Greenland ice sheet and West Antarctic ice sheet, tropical coral reef die off, and permafrost abrupt thaw
- International collaboration on climate models demonstrates a range of potential future scenarios that account for assumptions on population, land use, energy intensity, energy use, and regionally differentiated development. These models demonstrate the potential impacts of global warming under different mitigation and adaptation efforts

The Earth's natural greenhouse

The temperature of the Earth is determined by the balance between energy received from the Sun and its loss back into space (IPCC, 2021). The Sun's energy consists of short-wave radiation (mainly visible 'light' and ultraviolet (UV) radiation) and nearly all of it passes through the atmosphere without interference (see Figure 1.1). The only exception is damaging high-energy UV, which is absorbed by atmospheric ozone. About one-third of the solar energy is reflected straight back into space. The remaining energy is absorbed by the surface of the Earth. This energy warms the land and the oceans, and the heat is radiated back as long-wave infrared or 'heat' radiation. Atmospheric gases, such as water vapour, carbon dioxide (CO_2), methane (CH_4), and nitrous oxide (N_2O),

DOI: 10.4324/9781003545224-2

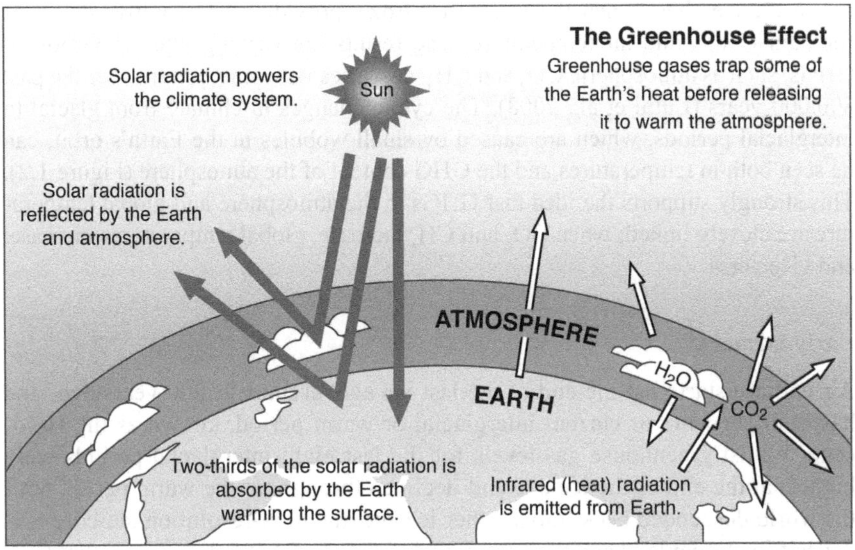

Figure 1.1 Explanation of the greenhouse effect. Based on a figure from IPCC (1992).

are known as greenhouse gases (GHGs) as they absorb some of this long-wave radiation, warming the atmosphere. This effect has been measured in the atmosphere and can be reproduced time and time again in the laboratory. Without this natural greenhouse effect, the Earth would be at least 35°C colder, making the average temperature in the tropics about −10°C (Maslin, 2021a). Since the industrial revolution, we have been burning fossil fuels (oil, coal, and natural gas) deposited hundreds of millions of years ago, releasing the carbon back into the atmosphere as CO_2 and CH_4, increasing the 'greenhouse effect,' and elevating the temperature of the Earth (Mann, 2021).

Past variations in carbon dioxide

Study of the past shows that atmospheric CO_2 is an important control of global climate. Evidence for past variations in GHGs and temperature comes from ice cores drilled in both Antarctica and Greenland. As snow falls, it is light and fluffy and contains a lot of air. When more snow falls on top, the older snow is slowly compacted to form ice, and some of these air bubbles are trapped. By extracting these air bubbles trapped in the ancient ice, scientists can measure the percentage of GHGs that were present in the past atmosphere. Scientists have drilled down over two miles into both the Greenland and Antarctic ice sheets, which has enabled them to reconstruct the amount of GHGs present in the atmosphere over the past million years. By examining the oxygen and hydrogen isotopes in the frozen water that make up the ice core, it is also possible to estimate the air temperature above the ice sheet when the water first froze, providing comparative data sets for both GHGs and air temperature. The results are striking: the proportion of GHGs, such as atmospheric CO_2 and CH_4, co-varies with temperature over the past 800,000 years (Lüthi et al., 2008). The cyclic changes in climate from glacial to interglacial periods, which are caused by small wobbles in the Earth's orbit, can be seen both in temperatures and the GHG content of the atmosphere (Figure 1.2). This strongly supports the idea that GHGs in the atmosphere and global temperature are closely linked; when CO_2 and CH_4 increase, global temperatures increase, and vice versa.

Early farmers

Agriculture started at the end of the last ice age, around 12,000 years ago, and has flourished in our current interglacial or warm period, known as the Holocene. Whilst greenhouse gas levels for the last eight interglacial periods were highest at the end of the ice age and declined throughout the warm period until the world descended back into another ice age, the high-resolution ice-core evidence from Greenland and the continental margins of Antarctica shows that Holocene GHGs rose a small amount before the industrial revolution in the 1700s. Ruddiman and colleagues (2020) suggested that early agriculturalists caused

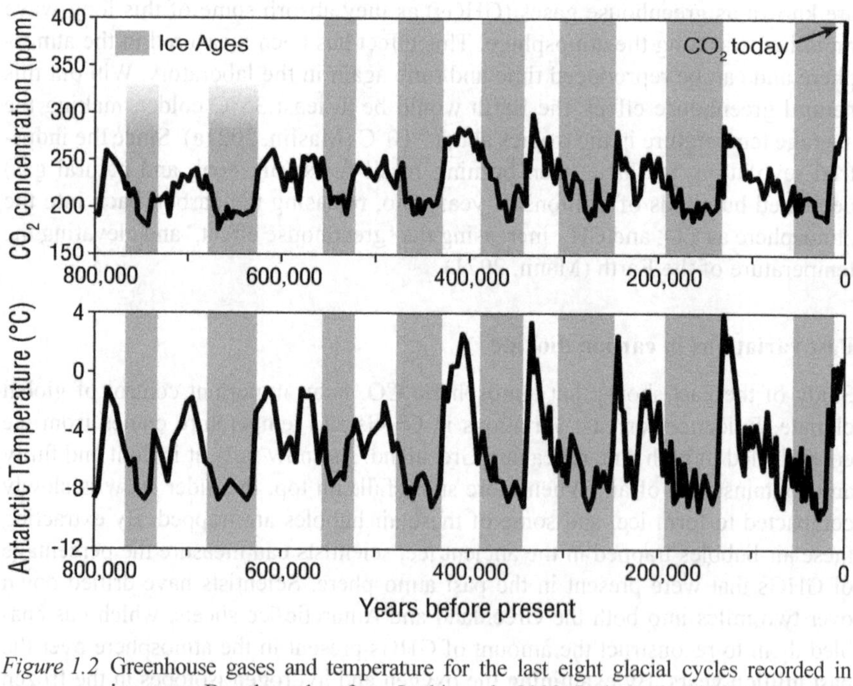

Figure 1.2 Greenhouse gases and temperature for the last eight glacial cycles recorded in ice cores. Based on data from Lüthi et al., 2008, Petit et al., 1999, and Shakun et al., 2012.

this reversal in natural decline in GHGs. Deforestation and land clearing led to a rise in atmospheric CO_2 (due to the removal of natural carbon sinks—reservoirs that absorb more CO_2 from the atmosphere than they release) starting about 7,000 years ago, while the expansion of wet rice agriculture and domestic cattle caused atmospheric CH_4[1] to start rising about 5,000 years ago. It seems that early human interactions with our environment increased atmospheric GHGs just enough that even prior to the industrial revolution we had already delayed the onset of the next ice age, which many experts think would have started in the near future.

The industrial revolution

There is clear evidence, from both ice core data and direct atmospheric measurements, that levels of atmospheric CO_2 have been rising ever since the beginning of the industrial revolution, in the late 19th century. The first direct measurements of CO_2 concentrations in the atmosphere started in 1958, on the summit of Mauna Loa Mountain in Hawaii at an altitude of about 4,000 m. The measurements were made at this remote location to avoid contamination from local pollution sources, and the record clearly shows that atmospheric concentrations of CO_2 have increased every single year since 1958. The mean concentration of approximately 316 parts per million by volume (ppm) in 1958 has risen to over 420 ppm today (see Figure 1.3).

The Mauna Loa observatory CO_2 data can be combined with the detailed ice-core evidence to produce a complete record of atmospheric CO_2 since the beginning of the industrial revolution. This shows that atmospheric CO_2 has increased from a pre-industrial concentration of about 280 ppm to over 430 ppm at present, representing an increase of over 50%. To put this increase into context, ice-core evidence shows that over the last 800,000 years, the natural change in atmospheric CO_2 has been between about 200 and 280 ppm. The variation between warm and cold periods is about 80 ppm—less than the CO_2 pollution that we have put into the atmosphere over the past 100 years. The level of human pollution caused in one century is greater than natural variations, which took thousands of years.

History of climate change

Scientists predict that if we continue on our current carbon emission pathway we could warm the planet by between 1.5°C and 4.7°C by the end of this century (IPCC, 2021), which economists suggest could cost us as much as 10% of global gross domestic product (GDP) (Waidelich et al., 2024) and up to 18% in some low-income countries (Jones, 2022). In the face of such a threat, it is essential to understand the history of climate change and the evidence that supports it. The essential science of climate change was available in the late 1950s, but it was not taken seriously until the late 1980s. Since then, climate change has emerged as one of the biggest scientific and political issues facing humanity.

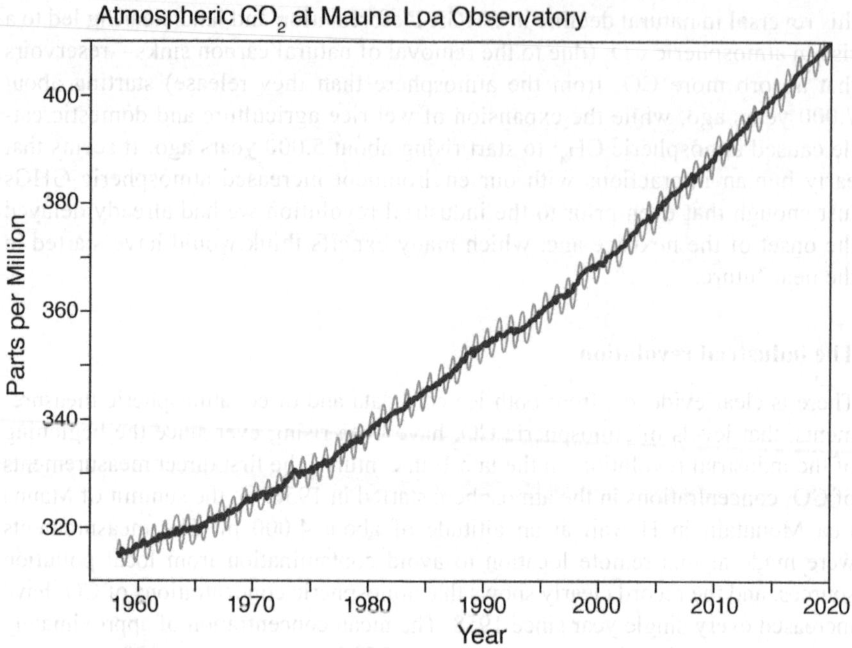

Figure 1.3 Mauna Loa observatory atmospheric carbon dioxide (CO_2) measurements. Based on open-access data from gml.noaa.gov/ccgg/trends/.

An old science

The history of climate change science is a long one and can be said to have started in 1856 when Eunice Newton Foote (American scientist, inventor, and women's rights campaigner) published her paper demonstrating the greenhouse effect of CO_2 (Maslin, 2021a). She used glass cylinders and mercury thermometers and showed that when they were filled with different gases and placed in direct sunlight, the one containing CO_2 trapped the most heat. Looking to the history of the Earth, Foote theorised: 'an atmosphere of that gas would give to our Earth a higher temperature' (Foote, 1856, p. 383).

Just three years later, John Tyndall, Professor of Natural Philosophy at the Royal Institution in London, demonstrated and measured the greenhouse effects of different gases (Bell, 2021). Using apparatus that converts thermal energy into electrical energy (thermopiles), he was the first to correctly measure the relative infrared (heat) absorption of gases such as nitrogen (N_2), oxygen (O_2), water vapour H_2O (g), CO_2, CH_4, and ozone (O_3). He concluded that H_2O (g) is the strongest absorber of radiant heat in the atmosphere and is the principal gas controlling air temperature of the Earth.

Building on the prior work of scientists such as Tyndall, Joseph Fourier, and Claude Pouillet, in 1896, the Swedish physical chemist Svante Arrhenius calculated how much the Earth's temperature would change due to variations in greenhouse gases (Bell, 2021). He estimated that a halving of atmospheric CO_2 would drop the Earth's temperature by 4°C and may have been a key cause of the ice ages, while a doubling of CO_2 would increase the global temperature by 4°C. He concluded that human-produced CO_2 emissions resulting from the burning of fossil fuels would be large enough to cause global warming (Bell, 2021).

But it was not until 1938 that the engineer and inventor Guy Stewart Callendar compiled 147 temperature records from around the world covering the previous 50 years, and showed that the world was, indeed, warming (Bell, 2021). Callendar's results were initially met with scepticism, but his papers, published through the 1940s and 1950s, prompted other scientists to investigate changes in atmospheric CO_2 and what was controlling them.

The Second World War saw a massive improvement in technology, including the development of infrared spectroscopy, and soon after the war, scientists showed that CO_2 in the upper atmosphere at low pressure did absorb heat—demonstrating the greenhouse effect (Weart, 2008). The worry about possible global warming was, however, ignored at first as scientists argued that any extra CO_2 emitted would simply be absorbed by the oceans. Roger Revelle, director of the Scripps Institute of Oceanography in California, was concerned by this dismissal (Weart, 2008). Through his studies of surface ocean chemistry, he found that oceans have a limited capacity for CO_2 absorption due to their chemical make-up, meaning they return much of the CO_2 that they absorb back into the atmosphere. This was a great revelation showing the oceans would not be the complete sink for anthropogenic CO_2, as was first thought. We now know that the oceans are taking up about one-quarter of the annual total anthropogenic production.

Charles Keeling, who was hired by Roger Revelle, made the next important step forward in climate change science. In the late 1950s and early 1960s, Keeling used the most modern technology available to measure the concentration of atmospheric CO_2 in Antarctica and on Mauna Loa. The resulting graph of daily CO_2 concentrations, known as the Keeling curve, has continued to climb ominously each year since his first measurement in 1958 and has become one of the major iconic images illustrating global warming (Figure 1.3).

Why the delay in recognising climate change?

In 1959, the physicist Gilbert Plass published an article in *Scientific American* declaring that the world's temperature would rise by 3°C by the end of the century. The magazine editors published an accompanying photograph of coal smoke belching from factories and the caption read, 'Man upsets the balance of natural processes by adding billions of tons of CO_2 to the atmosphere each year' (Plass, 1959). This resembles thousands of magazine articles, television news items, and documentaries that we have all seen since the late 1980s. So why was there a delay between the science of global warming being accepted in the late 1950s and the realisation by those outside the scientific community of the true threat of global warming at the beginning of the 21st century?

Maslin (2021a) argues that the key reasons for the delay in recognising climate change were the lack of increase in global temperatures and the lack of global environmental awareness. The global mean temperature (GMT) data set is calculated by compiling all the available land and sea temperatures. Despite a gradual increase in global temperatures since the industrial revolution, there appeared to be a slight downward trend from 1940 until the mid-1970s. This provoked many scientists to discuss whether the Earth was entering the next great ice age (Weart, 2008). Increasing knowledge of past climates in the 1970s and 1980s showed that this was highly unlikely, as ice ages take thousands of years to form. However, this slight decrease in global temperatures is now known to have been caused by high concentrations of sulphate aerosols in the atmosphere (Mitchell et al., 1995). It was not until the late 1980s, when the global annual mean temperature curve started to increase again, that the global cooling scenario was finally discarded (Weart, 2008). By the late 1980s, the global annual mean temperature curve was rising so steeply that all the dormant evidence from the late 1950s and 1960s was dusted off and the global warming theory came to prominence. In fact, it was in 1988 that Professor Jim Hansen, the director of the NASA Goddard Institute for Space Studies, was asked to testify on the matter before the United States Senate Committee on Energy and Natural Resources. He stated that, 'global warming has reached a level such that we can ascribe with a high degree of confidence a cause-and-effect relationship between the greenhouse effect and observed warming . . . It is already happening now' (Weart, 2008). This testimony was widely reported by the media, and global warming became a mainstream issue (Weart, 2008).

Even though the planet has been warming up for the last 150 years, it seems that the eventual recognition of climate change was driven by the upturn in the global annual mean temperature data set in the 1980s. The latest IPCC (2021) Assessment Report has reviewed and synthesised a wide range of data sets. It shows that the trend in global temperature is correct, and that this warming trend has continued unstopped until the present day.

The rise of the global environmental social movement

The 1980s saw a massive grassroots expansion in the environmental movement, particularly in the USA, Canada, and the UK, partly as a backlash against the right-wing governments of the 1980s and the expansion of the consumer economy, and partly because of the increasing number of environment-related stories in the media (Corfee-Morlot et al., 2007). This heralded a new era of global environmental awareness and the emergence of transnational non-governmental organisations (NGOs). The roots of this growing environmental awareness can be traced back to a number of key markers. These include:

- the publication of Rachel Carson's *Silent Spring* in 1962 (Carson, 1962)
- the image of Earth seen from the Moon in 1969
- the Club of Rome's 1972 report on *Limits to Growth* (Meadows et al., 1972)
- the Three Mile Island nuclear reactor accident in 1979

- the nuclear accident at Chernobyl in 1986
- the Exxon Valdez oil spillage in 1989

But these environmental problems were all regional in effect and limited geographically to the specific areas in which they occurred (Lewis & Maslin, 2018), while the events and publications targeted only parts of society.

It was the discovery in 1985 by the British Antarctic Survey of depletion of the ozone layer over Antarctica that demonstrated the global connectivity of our environment. Ozone (O_3) is a highly reactive gas made of three oxygen atoms, and in the upper atmosphere, it absorbs harmful ultraviolet radiation coming from the Sun—which can cause cancer and damage DNA. The ozone 'hole' also had a tangible international cause, the use of chlorofluorocarbons (CFCs), which led to a whole new area of politics: the international management of the environment and atmospheric conditions. There followed a set of key international agreements signed by the leaders of the world, including the 1985 Vienna Convention for the Protection of the Ozone Layer, the 1987 Montreal Protocol on Substances that Deplete the Ozone Layer, and the 1990 London and 1992 Copenhagen Adjustments and Amendments to the Protocol. These have been held up as examples of successful environmental diplomacy (Maslin, 2021a) and explain why it was initially assumed it would be straightforward to have similar agreements for carbon emissions.

These new global environmental concerns and the ability to deal with them internationally were encouraged and articulated by the leading politicians at the time. Margaret Thatcher, the Prime Minister of the UK in 1989, gave an address to the UN outlining the science of climate change, the threat it posed to all nations, and the action needed to avert the crisis. She summed up by saying, 'We should work through this great organisation and its agencies to secure world-wide agreements on ways to cope with the effects of climate change, the thinning of the ozone layer, and the loss of precious species' (Leggett, 2018). George Bush Senior, President of the United States, gave similar speeches, including one in 1992 when he outlined his clear skies and global climate change initiatives at the National Oceanic and Atmospheric Administration (Leggett, 2018).

The Intergovernmental Panel on Climate Change (IPCC) was set up in 1988 and produced its very first science report in 1990. Two years later, with support from leaders from all around the world, the UN held the Rio Earth Summit, officially called the United Nations Conference on Environment and Development (UNCED), to help member states cooperate on sustainability and protect the world's environment. The Summit was a huge success and led to the *Rio Declaration on Environment and Development,* the local sustainability initiative called *Agenda 21,* and *Forest Principles.* It also set up the *United Nations Convention to Combat Desertification,* the *Convention on Biological Diversity* and the *Framework Convention on Climate Change (UNFCCC)* that underlies global negotiations to limit global greenhouse gas emissions (see Chapter 3). The Rio Earth Summit also laid the foundations for the United Nations Millennium Development Goals and the subsequent United Nations Sustainable Development Goals (SDGs) (Lewis & Maslin, 2018).

Economists calculate the impacts of climate change

Economists have been involved with studying climate change from the very beginning of the IPCC process in the late 1980s. One of the most important reports was the publication of the UK government-commissioned Stern Report on *The economics of climate change* published in 2006. The report was led by Sir Nicholas Stern, the then-advisor to the UK government on the economics of climate change and development reporting to Prime Minister Tony Blair. The report states that if we do nothing, the impacts of climate change could cost between 5% and 20% of global GDP every year. That means the whole world loses one-fifth of what it earns to address the impacts. But the Stern (2006) Report does present some good news, because it argued that if we do everything we can to reduce global GHG emissions and ensure we adapt to the coming effects of climate change, this will cost us only 1% of global GDP every year.

The Stern Report was criticised by other economists—for example, does it use the right inherent discount rate? This is the rate economists use to account for the fact that consumption inherently has a lower value in the future than in the present. In other words, future consumption should be discounted simply because it takes place in the future and people generally prefer the present to the future. For example, the Nobel laureate William Nordhaus used inherent discount rates of up to 4.3% (Nordhaus, 2007), which means that people today value environmental benefit occurring 25 years in the future half as much as the same benefit today. Nordhaus, however, has recently come under intense criticism as he claims a 4°C increase in global temperature over pre-industrial levels would only reduce GDP per capita by between 2% and 4%. But the fundamental flaw in this model is that he underestimates the costs of climate impacts, which were assumed to be only marginal—so even catastrophic levels of climate change do not do much economic damage in this economic model. The Stern Report has also been criticised for being overly optimistic about the costs of adapting to a low-carbon world. In June 2008, Sir Nicholas Stern did revise his estimated costs up to 2% of global GDP to shift to a carbon-free global economy (Stern, 2008). Nevertheless, the Stern Report sent seismic waves around the world. It was as if people said to themselves, 'If the economists are worried about the cost of climate change, it must be real.'

This was not the end of the involvement of economists in climate change, and there have been a number of highly influential books and papers which question our whole understanding of economics and its relationship with the environment. These include the economist Tim Jackson's book *Prosperity Without Growth,* first published in 2009, that challenges the orthodox view that economic growth is required or even desirable (Jackson, 2016). In 2017, economist Kate Raworth published *Doughnut Economics,* in which she shows seven ways classical economics has got it wrong, arguing that environmental limits and basic human rights must be at the centre of economics (Raworth, 2017). For the first time in two generations, 'classical economics' is under sustained attack from a new generation of dynamic, innovative, 21st-century economists who see environmental and human well-being as integral parts of the world economy. At the centre of this is how we deal with climate change while improving peoples' lives (see Chapter 5 on the role of purpose as being founded in the notion of the well-being economy).

Climate change and the media

The other reason for the emergence of climate change as a major global issue has been the intense media interest. This is because climate change is perfect for the media: a dramatic story about the end of the world as we know it, with some key protagonists on the other hand arguing that it is not even real. The majority of newspaper articles in the UK, the USA, and Australia in the 1990s cast doubt on the claims of climate change. There was a recurrent attempt to promote mistrust in science, through strategies of generalisation, exaggerating disagreement within the scientific community, and, most importantly, discrediting scientists and scientific institutions (Oreskes & Conway, 2012).

There are two possible explanations for this extraordinary media-facilitated public scientific debate. First, climate change deniers and industrial lobby groups who do not want to see political action to address climate change use this debate about methods and scientific uncertainty as a convenient hook on which to hang their case for delay. As noted in the Introduction, oil and gas companies in the US have reportedly spent hundreds of millions over successive years, including $124.4 million in 2022 to lobby central government to control, delay, or block binding climate policy (McCarthy, 2019; Sayki & Cloutier, 2023).

Second, the media's ethical commitment to balanced reporting, inappropriately applied, draws unwarranted attention to critical views when they are marginal and outside the realm of what is normally considered 'good' science. In the UK, the BBC came under increasing criticism for continually presenting this false balance, usually pitching a climate scientist against a seasoned politician or a paid lobbyist.

Beyond conventional media, the so-called debate about climate change has now moved on to social media, with climate change deniers attacking the evidence and views of scientists whenever they can. This rise of fake news has impacted many areas of science including vaccinations and efforts to tackle Covid-19, as well as climate change. Together, false balance in media debate, fake news, and social media campaigns contribute to a public impression that the science of climate change is 'contested,' despite what many would argue is an overwhelming scientific case that climate change is occurring, and that human activity is its main driver. Studies have shown that there is 99% consensus on human-caused climate change in the peer-reviewed scientific literature (Lynas et al., 2021).

But things are changing, and in the past few years in many countries, opinion polls have shown that the majority of the public have realised that climate change is real and a major threat—this is despite the recent resurgence of anti-climate change rhetoric, particularly in the context of political movements in the US (Singh Khadka, 2025). Nonetheless, the move towards broader global acceptance of the climate change evidence is arising mainly through people's own experience or watching the effects of extreme weather around the world. There are now regular news stories about climate change and major documentaries such as Al Gore's 'An Inconvenient

Truth' (2006), David Attenborough's 'A Life on Our Planet' (2020), and BBC 1's 'Climate Change: The Facts' have also drawn widespread attention to the issue.

The new global environmental social movements

In May 2018, the protest group Extinction Rebellion was set up in the UK with the aim to use non-violent civil disobedience, backed by over 100 academics, to compel Governments to take action on climate change. In November 2018 and April 2019, they brought central London to a standstill by blockading major roads and bridges. Extinction Rebellion then spread to another 60 other cities around the world.

In August 2018, the 15-year-old student Greta Thunberg started to spend her school days outside the Swedish parliament holding a sign saying *Skolstrejk för klimatet* (school strike for climate), calling for stronger action on climate change. The message spread. Soon, other students all around the world started similar school strikes once a month on a Friday and they called the movement 'Fridays for Future.' It has been estimated that by the end of 2019 there had been over 4500 strikes across more than 150 countries, involving four million school children.

In 2018 and 2019, three extremely influential IPCC reports were published. The first, in 2018, was the Special Report on Global Warming of 1.5°C (IPCC, 2018), which documented what the world needed to do to keep global temperature rise below 1.5°C. It also showed that such significant mitigation actions can interact with, and often positively support, the achievement of multiple Sustainable Development Goals (SDGs)—a set of 17 interconnected global goals adopted by the United Nations—though some trade-offs may exist. The second was a special report on Climate Change and Land (IPCC, 2019a) and showed how climate change would impact desertification, land management, food security, and the terrestrial ecosystems. The third was the IPCC Special Report on the Ocean and Cryosphere (IPCC, 2019b), showing the impacts of climate change on the speed of melting of ice sheets, mountain glaciers, and sea ice, and their effects on sea level rise and marine ecosystems.

This new social movement and the very latest science have inspired many corporations to take a leading role. For example, Microsoft has set the agenda for the technology sector with the ambitious target to go carbon negative by 2030. By 2050, the company aims to remove all the carbon pollution from the atmosphere that they and their supply chain have emitted since the founding of the company in 1975. Sky has set the agenda for the media sector, pledging that they and their supply chain will go carbon negative by 2030. These companies form part of a group of over 4,000 global companies that have pledged to adopt science-based targets to all be net-zero carbon emitters by 2050 (https://sciencebasedtargets.org/).

Scientific evidence for climate change

Science is not a belief system. It is a rational, logical methodology that moves forward by using detailed observation and experiments to constantly test and retest ideas and theories. It is the very foundation of our global society. Over the last 150 years, a huge amount of scientific evidence has been collected showing how the climate system operates and that humans are now causing it to rapidly change.

Weight of evidence

If we are to understand climate change, we must understand how science works. The 'weight of evidence' principle prompts the constant need to compile new data and undertake new experiments in order to continually test our ideas and theories regarding climate. Over the past 40 years, the theory of climate change must have been one of the most comprehensively tested ideas in science. There are six main areas of evidence that should be considered:

1. We have tracked the rise in GHGs in the atmosphere. Carbon dioxide has risen by over 50% and CH_4 by 150% since pre-industrial times. We also understand the role of greenhouse gases in driving warming and cooling of past climates.
2. We know from laboratory and atmospheric measurements that GHGs do indeed absorb heat when they are present in the atmosphere. Table 1.1 summarises the latest understanding of the main GHGs.
3. We have tracked significant changes in the climate system over the last century. Temperatures have risen by over 1.5°C and sea levels by over 24 cm since the beginning of the 20th century. Between 1880 and 2024, the ten warmest years on record have all occurred in the past ten years (NOAA, 2024). 2024 was the warmest year ever recorded. At the time of writing, in the last 12 months, global temperatures have crossed the 1.5°C threshold set by global leaders in Paris in 2015. Our oceans are warming, which means they can hold less dissolved gases, and they have already lost over 2% of their dissolved oxygen, which is essential to marine life.
4. We have analysed the effects of other natural changes on climate, including sunspots and volcanic eruptions, and though these are essential to understanding the pattern of temperature changes over the past 150 years, they cannot explain the current warming trend.
5. We have observed significant changes in the Earth system (Ripple et al., 2023), including: melting of the Greenland and Western Antarctic ice sheets; retreating of Arctic sea ice; retreating mountain glaciers on all continents; shrinking of the permafrost regions; changes in seasonal patterns of plant growth; changes in bird migration patterns; and shifts in the geographic ranges of some plants and animals.
6. We continually track global weather, which is changing. Extreme weather events have increased all around the world (Ripple et al., 2023), including: super-storms, mega-floods, severe droughts, unprecedented heatwaves, and uncontrollable wildfires. Attribution studies demonstrate that climate change has made a significant contribution to the increase in extreme weather events (Clarke et al., 2022).

Table 1.1 Main greenhouse gases and their comparative ability to warm the atmosphere.

Greenhouse gas	*Chemical formula*	*Lifetime** *(years)*	*Pre-industrial levels*	*2024 levels*	*Human source*	*Global warming potential***	
						20 yrs	**100 yrs**
Carbon dioxide	CO_2	***	278 ppm	422 ppm (>50% increase)	Fossil-fuel combustion Land-use changes Cement production	1	1
Methane	CH_4	12.4	700 ppb	1943 ppb (>175% increase)	Fossil-fuels Rice paddies Waste dumps Livestock	96	32
Nitrous oxide	N_2O	121	275 ppb	338 ppb (~25% increase)	Fertiliser Industrial processes Fossil-fuel combustion	264	265
CFC-12	CCl_2F_2	100	Not naturally occurring	508 ppt	Liquid coolants/foams	10,800	10,200
HCFC-22	$CHClF_2$	11.9	Not naturally occurring	244 ppt	Liquid coolants	5,280	1,760
Perfluoro methane PCF-14	CF_4	50,000	Trace	79 ppt	Production of aluminium	4,880	6,630
Sulphur hexa-fluoride	SF6	3,200	Trace	9.59 ppt	Dielectric fluid	17,500	23,500

ppm = parts per million in the atmosphere
ppb = parts per billion in the atmosphere
ppt = parts per trillion in the atmosphere

* Lifetime refers to the time it takes the gas to decay to approximately by 36.8% of its original concentration, crucial for understanding how quickly chemical reactions occur and how long substances persist in the atmosphere. For instance, atmospheric lifetimes of greenhouse gases influence their global warming potential and how long they remain in the atmosphere

** Global warming potential (GWP) quantifies how much a greenhouse gas contributes to warming the planet compared to CO_2 over a specific time period. CO_2 is used as the reference, with a GWP of 1. Other gases have GWPs that indicate how many times more warming they cause than the same amount of CO_2

*** The longevity of CO_2 in the atmosphere is complex. 60–80% would be absorbed relatively quickly on a timescale of 200–2000 years, whilst 20–40% would remain in the atmosphere awaiting slower reactions (Archer et al., 2009)

Modelling future climate

The whole of human society operates on knowing the future weather. For example, farmers in India know when the monsoon rains will come next year and so they know when to plant the crops. Farmers in Indonesia know there are two monsoon rains each year, so next year they can have two harvests. This is based on their knowledge of the past, as the monsoons have always come at about the same time each year in living memory. But the need to predict goes deeper than this; it influences every part of our lives. Our houses, roads, railways, airports, offices, cars, trains, and so on are all designed for the local climate. For example, in England, all the houses have central heating as the outside temperature is usually below 20°C, but no air-conditioning, as temperatures rarely exceed 26°C, while in Australia, the opposite is true: most houses have air-conditioning but rarely central heating. Predicting future climate is now essential, as we can no longer rely on records of past weather to tell us what the future will hold. We also need to understand the consequences of our actions. For example, if we continue to emit GHGs at the same rate as today, how much climate change will occur? So, we must develop new ways of understanding potential futures by modelling.

There is a whole hierarchy of climate models, from relatively simple box models to extremely complex three-dimensional General Circulation Models (GCMs). Each has a role in examining and furthering our understanding of the global climate system (Maslin, 2021a). It is the complex three-dimensional GCMs that are used to predict future global climate. These comprehensive climate models are based on physical laws represented by mathematical equations, which are solved using a three-dimensional grid over the globe. To obtain the most realistic simulations, all the major parts of the climate system must be represented in sub-models, including atmosphere, ocean, land surface (topography), cryosphere, and biosphere, as well as the processes that go on within them and between them.

Carbon cycle

At the heart of the climate models is the carbon cycle, which is central to estimating what happens to anthropogenic CO_2 and CH_4 emissions. The Earth's carbon cycle is complicated, with both large sources and sinks of CO_2. Currently, half of all our carbon emissions are absorbed by the natural carbon cycle and do not end up in the atmosphere, but rather in the oceans and the terrestrial biosphere. Figure 1.4 shows the global carbon reservoirs in gigatonnes (GtC, or 1,000 million tonnes) and fluxes (the movement and exchange of carbon in GtC per year). These indicative figures show the changes since the industrial revolution. What has become clear is that the ocean surface and the land biosphere both take up about 25% each of our carbon emission every year. As the oceans continue to warm, however, they can hold less dissolved CO_2, which means their uptake will reduce. As we continue to deforest and substantially alter land use, the ability of the land biosphere to absorb carbon diminishes.

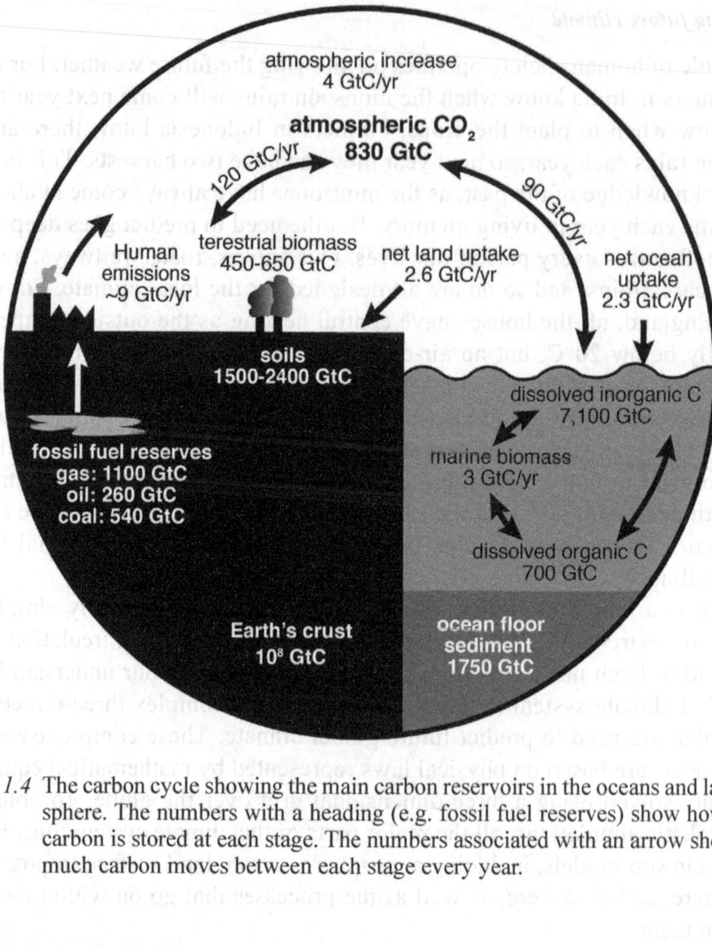

Figure 1.4 The carbon cycle showing the main carbon reservoirs in the oceans and land biosphere. The numbers with a heading (e.g. fossil fuel reserves) show how much carbon is stored at each stage. The numbers associated with an arrow show how much carbon moves between each stage every year.

Warming and cooling effects

As well as the warming effects of the GHGs, the Earth's climate system is complicated in that there are also cooling effects. This includes the amount of aerosols (tiny solid or liquid particles) in the air, which have a direct effect on the proportion of solar radiation that hits the Earth's surface. These aerosols can be both natural, like sea salt or dust, or caused by human activities, such as sulphate from industrial sulphur dioxide emissions or pollution particles from traffic emissions. Computer simulations of climate change demonstrate that industrial areas of the planet have *not* warmed as much as would be predicted just from rising GHGs. This so-called 'global dimming,' or more precisely 'regional dimming,' has been confirmed with real temperature and aerosol measurements. Increasing aerosols in the atmosphere also increases the amount of cloud cover, as they provide points on which the water vapour can nucleate, meaning it can begin to form tiny droplets of liquid water. Whilst water vapour is a GHG, and hence contributes to global warming, the upper white surface of clouds also reflects solar radiation back into space. The level of reflectivity of a surface is called 'albedo.' Clouds and ice have a high albedo, which means that they reflect large quantities of solar radiation away from surfaces on Earth, and therefore increased cloud cover may have a cooling effect. Predicting what will happen to the amount and types of clouds, and their warming potential, has been one of the key challenges for climate scientists.

Emission models of the future

The biggest unknown in the models is not the physics, the chemistry, nor the biology; it is the estimation of future global GHG emissions over the next 80 years. This includes many variables, from the global economy to personal lifestyles. Individual models are therefore run many times with different emission scenarios to provide a range of changes that could occur in the future. In fact, the latest IPCC Sixth Assessment Report (AR6) has compiled the results of multiple runs from over 100 distinct climate models being produced across 49 different international modelling groups, which are all part of the Coupled Model Intercomparison Projects (CMIP6). Of course, as computer processing power continues to increase, both the representation of coupled climate systems and the spatial scale will continue to improve.

The future scenarios we currently have available are influenced by predicted or modelled population growth, economic growth, development, fossil-fuel usage, the rate at which we switch to alternative energy, the rate of deforestation, and the effectiveness of international agreements to cut emissions. Out of all the systems that we are trying to model into the future, humanity is by far the most complicated and unpredictable.

If you want to understand the problem of predicting what will happen in the next 80 years, imagine yourself in 1920 and think about what you would have predicted the world to be like in the 21st century. At the beginning of the 20th century, the British Empire was the dominant world power due to the industrial revolution and the use of coal. Would you have predicted the switch to a global economy based on oil after the Second World War? Or the explosion in car use? Or the general availability of air travel? Even 30 years ago, it would have been difficult to predict that there would be cheap and widespread access to air travel throughout Europe, the USA, and Asia.

In order to develop future scenarios of GHGs, it is essential that we know where our current GHG emissions are coming from. According to the IPCC, the primary source of CO_2 is the burning of fossil fuels; over 85% of global CO_2 emissions come from energy production, industrial processes, and transport. These are not evenly distributed around the world because of the unequal distribution of industry and wealth—North America, Europe, and Asia emit over 90% of the global, industrially produced CO_2. Moreover, historically, the developed nations have emitted much more than have the less developed countries.

The second major source, accounting for 10–15% of global CO_2 emissions, is land-use changes. These emissions come primarily from the cutting down of forests for the purposes of agriculture, urbanisation, or roads. When rainforest is cut down, the land often turns into less productive grassland with considerably reduced capacity for storing CO_2. Here, the pattern of CO_2 emissions is different, with South America, Asia, and Africa being responsible for over 90% of present-day land-use change emissions.

The first IPCC reports used simplistic assumptions of GHG emissions over the next 100 years. From 2000 onwards, climate models began to use more detailed scenarios, and the 2013 IPCC Fifth Assessment Report introduced Representative Concentration Pathways (RCPs), which considered a much wider set of inputs to the socio-economic models, including population, land use, energy intensity, energy use, and regional differentiated development. The RCPs were defined by the final radiative forcing achieved by the year 2100. Radiative forcing is defined as the difference between sunlight (radiant energy) received by the Earth and the energy radiated back to space and is measured in units of W/m^2 of the Earth's surface. Initially, the RCPs ranged from RCP2.6 to RCP8.5, signifying a range of final W/m^2 from 2.6 to 8.5. For the 2021 IPCC Sixth Assessment Report, an additional RCP (RCP1.9) was added to represent the Paris 2015 agreement ambition to keep global temperature rise to only 1.5˚C above pre-industrial levels.

The IPCC Sixth Assessment report also introduced Shared Socioeconomic Pathways (SSPs), which were developed to cover the full range of possible futures which then determine the RCPs at the regional scale. The SSPs are a set of five narratives and driving forces that may shape the global economy and global emissions in the future. The SSPs were defined by Riahi et al. (2017) as set out in Table 1.2.

The scenarios used in the Sixth Assessment Report are a combination of the SSPs and RCPs so that each SSP has seven associated RCPs totalling 35 scenarios (Table 1.3). This is because the SSPs do not include any mitigation measures, so a high-emission global economy could have a lower concentration pathway by employing huge amounts of mitigation, such as SSP5–1.9.

If all the SSP and RCP combinations are examined, then some are extremely unlikely, and some are almost impossible, for example, all of the scenarios under SSP5 and RCP1.9. The Sixth Assessment Report focuses on five main scenarios, combining SSPs and RCPs—SSP1–1.9, SSP1–2.6, SSP2–4.5, SSP3–7.0, and SSP5–8.5 (Table 1.3). Over 35 climate models have been run for all the different future scenarios discussed earlier, ranging from no climate action to doing everything to keep temperatures below 1.5˚C. These climate models suggest that, depending on how much GHGs we emit, the global surface temperature could, by the end of the century, rise between 1.5˚C and 5.5˚C compared with the preindustrial period. Figure 1.5 illustrates a range of potential temperature increases based on a sample of scenarios that achieve only 1.5°C and 2°C warming, and the existing national commitments (NDCs)[2] to climate change policy.

Tipping points

The term 'tipping points' is used a lot in climate change research and discussions. However, care must be taken, as there are two usages of this word. First, there are references to *climate* tipping points, which are the large-scale, often irreversible

Table 1.2 Shared socio-economic pathways representing projected socio-economic global changes up to 2100, the challenges they pose to mitigation and adaptation, and levels of optimism for human development.

SSP	*Mitigation versus adaptation*	*Optimism*	*Narrative*
SSP1: Sustainability	Low challenges to mitigation and adaptation	High	A shift towards environmental sustainability. Substantial investments in education and health, rapid economic growth, and well-functioning institutions.
SSP2: Middle of the road	Medium challenges to mitigation and adaptation	Medium	Continuation of historical trends. Development and income growth proceed unevenly. Environmental systems are facing a certain degradation.
SSP3: Regional rivalry	High challenges to mitigation and adaptation	Low	Growing international tensions, public services collapse. Investments in education and technological development decrease. Some regions suffer drastic environmental damage.
SSP4: Inequality	Low challenges to mitigation, but high challenges to adaptation	Low	Highly unequal investments in human capital, increasing inequality. Investment in both carbon-intensive and renewable energy sources. Environmental policies are localised.
SSP5: Fossil-fuelled development	High challenges to mitigation, but low challenges to adaptation	High	Social and economic development is based on an intensified exploitation of fossil fuel resources. Global economic growth, local environmental amelioration.

Table 1.3 Combinations of RCPs and SSPs showing the five main scenarios that were the focus in the latest IPCC Assessment Report (AR6).

2100 forcing level (W/m^2)	*SSP1 Sustainability*	*SSP2 Middle of the Road*	*SSP3 Regional rivalry*	*SSP4 Inequality*	*SSP5 Fossil fuelled development*
8.5					[shaded]
7.0			[shaded]		
6.0					
4.5		[shaded]			
3.4					
2.6	[shaded]				
1.9	[shaded]				

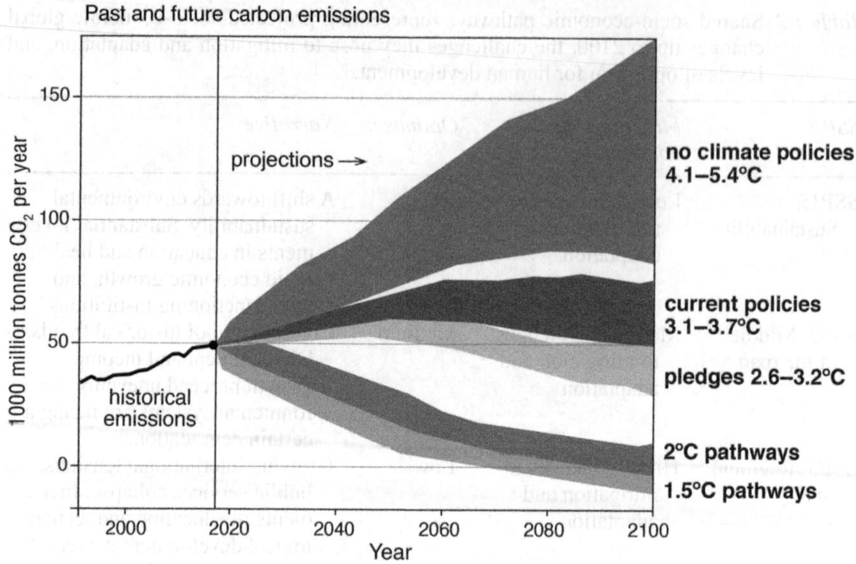

Figure 1.5 Future carbon emission scenarios and potential global warming by 2100 based on the scenarios which achieve a 1.5°C and 2°C pathway, the national policies and 'pledges' (NDCs) that were submitted at COP26 in Glasgow, and a scenario of no climate policies. Data source: Emissions data from Climate Action Tracker (Climate Action Tracker, n.d.).

shifts in the climate system, such as irreversible melting of ice sheets or the release of huge stores of CH_4 from below the oceans.

The other usage concerns *societal* tipping points, which occur when climate change has a major effect on a region or a particular country. For example, a 200-mile (~322-km) shift northward of the Southeast Asian monsoonal rainfall belt is, in climatological terms, a small shift, and is not a fundamental climate tipping point. But for the countries where the rains no longer fall or those where it does for the first time, such a shift is a major societal climate tipping point, because their weather may have been permanently altered (Maslin, 2021a).

Scientists have identified nine global core tipping points, which could lead to irreversible changes across the entire planet and seven regional impact climate tipping points, which could cause significant and potentially irreversible changes in those areas (Armstrong McKay et al., 2022).

Global core tipping points

- Greenland ice sheet collapse
- Western Antarctic ice sheet collapse
- Labrador Seas deep-water circulation collapse
- East Antarctic subglacial basins collapse

- Arctic Winter Sea Ice collapse
- Eastern Antarctic ice sheet collapse
- Amazon rainforest dieback
- High latitude permafrost collapse
- Atlantic deep-water circulation collapse

The regional tipping points include:

- Low-latitude Coral Reefs die off
- Abrupt thaw of the Boreal Permafrost
- Loss of Barents Sea ice
- Loss of mountain glaciers
- Greening of the Sahel and West Africa
- Boreal forest northern expansion
- Boreal forest southern die off

Out of these, Armstrong McKay and colleagues (2022) suggest that one regional and three global climate tipping points will likely pass the critical threshold if global warming stays consistently above 1.5°C. These are the Greenland ice sheet and West Antarctic ice sheet collapse, which combined could contribute more than 12 metres of global sea level rise, tropical coral reef die off, causing irreversible biodiversity loss, and boreal permafrost abrupt thaw, which could release vast amounts of stored CO_2 and CH_4. The other tipping points will occur beyond 2°C of global warming.

Summary: Year 2100 the nightmare scenario (SSP5–8.5)

If the 21st century draws to a close without significant action having been taken to prevent climate change, this is what we could expect (Maslin, 2021b).

Global temperatures have risen by over 4°C. In many countries, summer temperatures persistently stay above 40°C. Heatwaves with temperatures as high as 50°C have become common in tropical countries. Every summer, wildfires rage across every continent except Antarctica, creating plumes of acrid smoke that make breathing outdoors unbearable, causing an annual health crisis.

Ocean temperatures have risen dramatically. After repeated bleaching events, Australia's Great Barrier Reef has been officially declared dead, decimating fish populations. Frequent and prolonged droughts torment vast swathes of the Earth's land. The deserts of the world have expanded, displacing many millions of people. Around 3.5 billion people live in areas where water demand exceeds what is available. Air pollution has a new major cause outside the traffic-choked cities: dust whipped up from now-barren farmland.

The Arctic is free of sea ice every summer. Average temperatures in the far north have risen by over 8°C as a result. The Greenland and Western Antarctic ice sheets have started to melt, releasing a huge amount of freshwater into the oceans. Most mountain glaciers have completely melted. Skiing is now a predominantly indoor sport which takes place on giant artificial slopes. Most of the Himalayan plateau's ice has disappeared, reducing the flows of the Indus, Ganges, Brahmaputra, and Yamuna rivers, which over 600 million people rely on for plentiful water.

The extra heat in the ocean has caused it to expand. Combined with water from melting ice sheets, sea levels have risen by more than 1m. Many major cities, including Hong Kong, Rio de Janeiro, and Miami, are already flooded and uninhabitable. The Maldives, the Marshall Islands, Tuvalu, and many other small island nations have been abandoned as they are now under water. Many coastal and river areas are regularly flooded, including the Nile Delta, the Rhine valley, and Thailand. Over 20% of Bangladesh is permanently under water.

Winter storms are more energetic and unleash more water, causing widespread wind damage and flooding each year. Tropical cyclones have become stronger and affect tens of millions of people every year. Mega-cyclones, like 2013's Typhoon Haiyan, have become more common, with sustained wind speeds of over 200 mph. Southeast Asian monsoons have become more intense and unpredictable, bringing either too much or too little rain to each region, affecting the lives of over three billion people.

Food and water insecurity have increased around the world, threatening the health and well-being of billions of people. Extreme heat and humidity in the tropics and subtropics have increased the number of days that it is impossible to work outside tenfold, slashing farm productivity. Extreme weather in temperate regions like Europe has made food production highly unpredictable. Half of the land devoted to agriculture in the past is now unusable, and the capacity of the rest to grow food differs widely from season to season. Crop yields are at their lowest levels since the middle of the 20th century. Fish stocks have collapsed. The acidity of the ocean has increased by 125%. The ocean food chain has collapsed in some regions as the small marine organisms that form its base struggle to make calcium carbonate shells in the more acidic waters.

Despite advances in medical sciences, deaths from tuberculosis, malaria, cholera, diarrhoea, and respiratory illnesses are at their highest levels in human history. Extreme weather events—from heat-waves and droughts to storms and floods—are causing large loss of life and leaving millions of people homeless. Disease epidemics have plagued the century, spreading among populations beleaguered by widespread poverty and vulnerability. Climate migrants have become a major global crisis as people try to escape rising seas, extreme weather events, diseases, and poverty.

Summary: Year 2100 humanity rises to the challenge (SSP1–1.9)

This is what our planet could look like according to Maslin (2021b) if we do everything in our power to contain climate change. Global temperatures rose to 1.5°C in the 2030s and remained there for the rest of the century. Fossil fuels have been replaced by renewable energy. Over a trillion trees have been planted, sucking CO_2 from the atmosphere. The air is cleaner than it has been since before the industrial revolution.

Cities have been restructured to provide all-electric public transport and vibrant green spaces. Many new buildings have a photoelectric skin which generates solar energy and green roofs which cool the cities, making them a more pleasant place to live. High-speed electric trains reaching 300 mph link many of the world's major cities. Intercontinental flights still run, using large and efficient planes running on synthetic kerosene that is made by combining water and CO_2 sucked directly from the atmosphere.

Global diets have shifted away from meat. Farming efficiency has greatly improved during the transition from industrial-scale meat production to plant-based sustenance, creating more land to rewild and reforest. Urban farms and vertical farms have developed to ensure high nutritious plant-based crops are available in towns and cities.

Half of the Earth is dedicated to restoring the natural biosphere and its ecological services. Elsewhere, fusion energy is finally set to work at scale providing unlimited clean energy for the people of the 22nd century.

Our future depends on what decisions are made today.

Conclusion

The challenge of climate change must be seen within the current dominant political and economic landscape. Only by understanding the fundamental societal and economic causes of carbon emissions can we hope to be able to build systems that can rapidly reduce them. The dominant narrative of the 21st century is the global transition to net-zero emissions—which means every company, business, organisation, and government will have to greatly reduce their greenhouse gas emissions and offset the rest to ensure the world reaches net zero by 2050. At the same time as we deal with climate change, we need to ensure that we also tackle other global challenges, such as global poverty/inequality, environmental degradation, and global security. Future policies and international agreements need to provide win-win solutions that deal with the biggest challenges facing humanity in the 21st century.

This chapter is based on Mark Maslin's books, How to Save Our Planet: The Facts (2021) and Climate Change: A Very Short Introduction, 4th edition (2021).

Notes

1 Manure and gastroenteric releases from livestock are a leading contributor to atmospheric CH_4 emissions, accounting for about 30% of emissions, whilst wet rice cultivation contributes around 12% of human-caused CH_4 emissions via flooded rice paddies which create ideal conditions for CH_4 -producing bacteria. Notably, CH_4 is a powerful greenhouse gas that is more than 25 times more potent than CO_2 at trapping heat in the atmosphere.

2 Nationally determined contributions (NDCs) reflect the efforts by each country to reduce national emissions and adapt to the impacts of climate change. The Paris Agreement (Article 4, paragraph 2) requires each Party to prepare, communicate, and maintain successive NDCs that it intends to achieve.

Further reading

Maslin, M. (2019a). 'The five corrupt pillars of climate change denial'. *The Conversation.* https://theconversation.com/the-five-corrupt-pillars-of-climate-change-denial-122893

Maslin, M. (2019b). 'Five climate change science misconceptions—debunked'. *The Conversation*. https://theconversation.com/five-climate-change-science-misconceptions-debunked-122570
Maslin, M. (2021a). *Climate change: A very short introduction*. Oxford University Press.
Maslin, M. (2021b). *How to save our planet: The facts*. Penguin Life.

References

Archer, D., Eby, M., Brovkin, V., Ridgwell, A., Cao, L., Mikolajewicz, U., Caldeira, K., Matsumoto, K., Munhoven, G., Montenegro, A., & Tokos, K. (2009). Atmospheric lifetime of fossil fuel carbon dioxide. *Annual Review of Earth and Planetary Sciences*, *37*(1), 117–134. https://doi.org/10.1146/annurev.earth.031208.100206

Armstrong McKay, D. I., Staal, A., Abrams, J. F., Winkelmann, R., Sakschewski, B., Loriani, S., Fetzer, I., Cornell, S. E., Rockström, J., & Lenton, T. M. (2022). Exceeding 1.5 C global warming could trigger multiple climate tipping points. *Science*, *377*(6611), eabn7950. https://doi.org/10.1126/science.abn7950

Attenborough, D. (2020). *A life on our planet: My witness statement and a vision for the future*. Random House.

Bell, A. (2021). *Our biggest experiment: An epic history of the climate crisis*. Catapult.

Carson, R. (1962). *Silent spring*. Penguin Books.

Clarke, B., Otto, F., Stuart-Smith, R., & Harrington, L. (2022). Extreme weather impacts of climate change: An attribution perspective. *Environmental Research: Climate*, *1*(1), 012001. https://doi.org/10.1088/2752-5295/ac6e7d

Climate Action Tracker. (n.d.). *CAT 2035 climate target update tracker*. https://climateactiontracker.org/climate-target-update-tracker-2035/

Corfee-Morlot, J., Maslin, M., & Burgess, J. (2007). Climate science in the public sphere. *Philosophical Transactions A of the Royal Society*, *365*(1860), 2741–2776. https://doi.org/10.1098/rsta.2007.2084

Foote, E. (1856). Circumstances affecting the heat of the sun's rays. *American Journal of Science and Arts*, *22*, 382–383.

Gore, A. (2006). *An inconvenient truth: The planetary emergency of global warming and what we can do about it*. Rodale.

IPCC. (1992). *Climate change: The 1990 and 1992 IPCC assessments*. WMO.

IPCC. (2018). *Global warming of 1.5°C. An IPCC special report on the impacts of global warming of 1.5°C above pre-industrial levels and related global greenhouse gas emission pathways, in the context of strengthening the global response to the threat of climate change, sustainable development, and efforts to eradicate poverty*. V. Masson-Delmotte, P. Zhai, H.-O. Pörtner, D. Roberts, J. Skea, P. R. Shukla, A. Pirani, W. Moufouma-Okia, C. Péan, R. Pidcock, S. Connors, J. B. R. Matthews, Y. Chen, X. Zhou, M. I. Gomis, E. Lonnoy, T. Maycock, M. Tignor, & T. Waterfield (Eds.). Cambridge University Press. https://doi.org/10.1017/9781009157940

IPCC. (2019a). *Special report on the ocean and cryosphere in a changing climate*. H.-O. Pörtner, D. C. Roberts, V. Masson-Delmotte, P. Zhai, M. Tignor, E. Poloczanska, K. Mintenbeck, A. Alegría, M. Nicolai, A. Okem, J. Petzold, B. Rama, & N. M. Weyer (Eds.). Cambridge University Press. https://doi.org/10.1017/9781009157964

IPCC. (2019b). *Climate change and land: An IPCC special report on climate change, desertification, land degradation, sustainable land management, food security, and greenhouse gas fluxes in terrestrial ecosystems*. P. R. Shukla, J. Skea, E. Calvo Buendia, V. Masson-Delmotte, H.-O. Pörtner, D. C. Roberts, P. Zhai, R. Slade, S. Connors, R. van Diemen, M. Ferrat, E. Haughey, S. Luz, S. Neogi, M. Pathak, J. Petzold, J. Portugal Pereira, P. Vyas, E. Huntley, . . . J. Malley (Eds.). https://doi.org/10.1017/9781009157988.001

IPCC. (2021). *Climate change 2021: The physical science basis. Contribution of working group I to the sixth assessment report of the intergovernmental panel on climate change*.

V. Masson-Delmotte, P. Zhai, A. Pirani, S. L. Connors, C. Péan, S. Berger, N. Caud, Y. Chen, L. Goldfarb, M. I. Gomis, M. Huang, K. Leitzell, E. Lonnoy, J. B. R. Matthews, T. K. Maycock, T. Waterfield, O. Yelekçi, R. Yu, & B. Zhou (Eds.). Cambridge University Press. https://doi.org/10.1017/9781009157896

Jackson, T. (2016). *Prosperity without growth: Foundations for the economy of tomorrow*. Routledge.

Jones, M. (2022). *Climate change putting 4% of global GDP at risk, new study estimates*. Reuters. https://www.reuters.com/world/climate-change-putting-4-global-gdp-risk-new-study-estimates-2022-04-27/

Leggett, J. K. (2018). *The winning of the carbon war: Power and politics on the front lines of climate and clean energy.* Crux Publishing.

Lewis, S. L., & Maslin, M. A. (2018). *The human planet: How humans caused the Anthropocene*. Penguin and Yale University Press.

Lüthi, D., Le Floch, M., Bereiter, B., Blunier, T., Barnola, J. M., Siegenthaler, U., Raynaud, D., Jouzel, J., Fischer, H., Kawamura, K., & Stocker, T. F. (2008). High-resolution carbon dioxide concentration record 650,000–800,000 years before present. *Nature*, *453*(7193), 379–382. https://doi.org/10.1038/nature06949

Lynas, M., Houlton, B. Z., & Perry, S. (2021). Greater than 99% consensus on human caused climate change in the peer-reviewed scientific literature. *Environmental Research Letters*, *16*, 11. https://doi.org/10.1088/1748-9326/ac2966

Mann, M. (2021). *The new climate war: The fight to take back our planet.* Public Affairs.

Maslin, M. (2021a). *Climate change: A very short introduction.* Oxford University Press.

Maslin, M. (2021b). *How to save our planet: The facts*. Penguin Life.

McCarthy, N. (2019). Oil and gas giants spend millions lobbying to block climate change policies. *Forbes Business*. https://www.forbes.com/sites/niallmccarthy/2019/03/25/oil-and-gas-giants-spend-millions-lobbying-to-block-climate-change-policies-infographic/

Meadows, D. H., Meadows, D. L., Randers, J., & Behrens III, W. W. (1972). *The limits to growth—Club of Rome*. Club of Rome.

Mitchell, J. F., Johns, T. C., Gregory, J. M., & Tett, S. F. B. (1995). Climate response to increasing levels of greenhouse gases and sulphate aerosols, *Nature*, *376*(6540), 501–504. https://doi.org/10.1038/376501a0

NOAA. (2024). *State of the climate*. https://www.noaa.gov/topic-tags/state-of-climate

Nordhaus, W. (2007). A review of the stern review on the economics of climate change, *Journal of Economic Literature, 45*, 686–702. https://doi.org/10.1257/jel.45.3.686

Oreskes, N., & Conway, M. (2012). *Merchants of doubt: How a handful of scientists obscured the truth on issues from tobacco smoke to global warming.* Bloomsbury.

Petit, J. R., Jouzel, J., Raynaud, D., Barkov, N. I., Barnola, J. M., Basile, I., Bender, M., Chappellaz, J., Davis, M., Delaygue, G., Delmotte, M., Kotlyakov, V. M., Legrand, M., Lipenkov, V. Y., Lorius, C., Pépin, L., Ritz, C., Saltzman, E., & Stiévenard, M. (1999). Climate and atmospheric history of the past 420,000 years from the Vostok ice core, Antarctica. *Nature*, *399*(6735), 429–436. https://doi.org/10.1038/20859

Plass, G. N. (1959). Carbon dioxide and climate. *Scientific American*, *201*(1), 41–47. https://www.jstor.org/stable/24940327

Raworth, K. (2017). Why it's time for Doughnut economics. *IPPR Progressive Review*, *24*(3), 216–222. https://doi.org/10.1111/newe.12058

Riahi, K., van Vuuren, D. P., Kriegler, E., Edmonds, J., O'Neill, B. C., Fujimori, S., Bauer, N., Calvin, K., Dellink, R., Fricko, O., Lutz, W., Popp, A., Cuaresma, J. C., Samir, K. C., Leimbach, M., Jiang, L., Kram, T., Rao, S., Emmerling, J., Ebi, K., Hasegawa, T., Havlik, P., Humpenöder, F., Da Silva, L. A., Smith, S., Stehfest, E., Bosetti, V., Eom, J., Gernaat, D., Masui, T., Rogelj, J., Strefler, J., Drouet, L., Krey, V., Luderer, G., Harmsen, M., Takahashi, M., Baumstark, L., Doelman, J. C., Kainuma, M., Klimont, Z., Marangoni, G., Lotze-Campen, H., Obersteiner, M., Tabeau, A., Tavoni, M. (2017). The shared socioeconomic pathways and their energy, land use, and greenhouse gas emissions implications:

An overview. *Global Environmental Change*, *42*, 153–168. https://doi.org/10.1016/j.gloenvcha.2016.05.009

Ripple, W. J., Wolf, C., Gregg, J. W., Rockström, J., Newsome, T. M., Law, B. E., Marques, L., Lenton, T. M., Xu, C., Huq, S., Simons, L., & King, S. D. A. (2023). The 2023 state of the climate report: Entering uncharted territory. *BioScience*, *73*(12), 841–850. https://doi.org/10.1093/biosci/biad080

Ruddiman, W. F., He, F., Vavrus, S. J., & Kutzbach, J. E. (2020). The early anthropogenic hypothesis: A review. *Quaternary Science Reviews*, *240*, 106386. https://doi.org/10.1016/j.quascirev.2020.106386

Sayki, I., & Cloutier, J. (2023, February 22). Oil and gas industry spent $124.4 million on federal lobbying amid record profits in 2022. *Open Secrets*. https://www.opensecrets.org/news/2023/02/oil-and-gas-industry-spent-124–4-million-on-federal-lobbying-amid-record-profits-in-2022/

Shakun, J. D., Clark, P. U., He, F., Marcott, S. A., Mix, A. C., Liu, Z., Otto-Bliesner, B., Schmittner, A., & Bard, E. (2012). Global warming preceded by increasing carbon dioxide concentrations during the last deglaciation. *Nature*, *484*(7392), 49–54. https://doi.org/10.1038/nature10915

Singh Khadka, N. (2025). How Trump's 'drill, baby, drill' pledge is affecting other countries. *BBC Climate*. https://www.bbc.co.uk/news/articles/ce857O9xdk46

Stern, N. (2006). *The economics of climate change: The stern review*. Cambridge University Press.

Stern, N. (2008). The economics of climate change. *American Economic Review*, *98*(2), 1–37. https://doi.org/10.1257/aer.98.2.1

Waidelich, P., Batibeniz, F., Rising, J., Kikstra, J. S., & Seneviratne, S. I. (2024). Climate damage projections beyond annual temperature. *Nature Climate Change*, *14*(6), 592–599. https://doi.org/10.1038/s41558-024-01990-8

Weart, S. R. (2008). *The discovery of global warming: Revised and expanded edition*. Harvard University Press.

2 The impacts of global climate change in the context of organisational responsibility

Anita Wreford and Katie Manning

Key points

- Rising global temperatures and increased frequency of extreme weather events such as wildfires, droughts, and flooding are radically altering ecosystems and depleting natural resources, jeopardising global food security, posing threats to human health, and intensifying competition over scarce resources and climate-induced displacement
- Infrastructure and transport networks are becoming increasingly vulnerable to extreme weather disruptions, which has knock-on effects throughout supply chains, leading to rising costs for both producers and consumers
- The socio-economic impacts of climate change are not felt equally across all communities. Despite contributing the least to greenhouse gas (GHG) concentrations, the most vulnerable populations are at greater risk to climate change effects due to lower adaptive capacity and limited financial resources
- The impacts of climate change are complex and cascade through socio-economic systems, making it essential to consider the broader, systemic effects of environmental change
- Both mitigation and adaptation to climate change offer socio-economic opportunities. Recognising these potential positive impacts is a key factor in decreasing the harmful effects of global warming

Introduction

Over the last 75 years, organisations have driven greenhouse gas emissions on a path of exponential increase. In his 2006 landmark report, Nicholas Stern (Stern, 2006) characterised global warming as an 'externality,' in other words, a damage caused to others due to market activity whose cost is not met by those who actually cause it. While Stern characterised climate change as the 'largest ever market failure,' others have countered that it represents precisely the opposite, that the market system is simply performing too well. Yet, during the expansion of Western capitalism from the 1950s to 2000, the impacts of climate change were often recorded as one-off events. There were few systematic studies of the changing frequency and intensity of climate change impacts.

DOI: 10.4324/9781003545224-3

Today, however, there is no doubt that climate change is real, occurring now, and already affecting our social and economic systems. In its most recent series of reports, the Intergovernmental Panel on Climate Change (IPCC) stated that human-induced climate change has caused widespread adverse impacts and related losses and damages to nature and people. Increasing climate risks are projected to exacerbate existing vulnerabilities and social inequalities and inequities (Pörtner et al., 2022). Recent studies project that the world economy is facing an income reduction of 19% within the next 26 years due to climate damages arising from the last 40 years of GHG emissions. This amounts to US$38 trillion per year of climate-related damages (Kotz et al., 2024). Furthermore, the World Meteorological Organization (WMO) reports that there has been a sevenfold increase in reported disaster losses from extreme weather since the 1970s, and approximately US$143 billion per year of these costs is attributable to human-induced climatic change (Newman & Noy, 2023).

While the focus of this chapter is on the impacts of climate change, it is important to acknowledge the role that effective adaptation will play in reducing or even avoiding some of these altogether. And, ultimately, reducing emissions (mitigation) is the most important tool in avoiding the worst impacts of climate change. The first part of this chapter will therefore focus on the ways in which organisations have contributed to climate change, and how both gradual environmental changes and increased extreme weather events are impacting ecosystems, natural resources, and the human activities that rely upon them. The second part of this chapter is devoted to considering the opportunities for innovation and green growth and the potential for positive impacts arising from climate action.

Organisational drivers of climate change

Economic growth and the exploitation of nature have had a long and prosperous relationship. From the disappearance of the Easter Island civilisation to the Maya collapse, there are numerous examples of civilisations expanding, growing their economy, and eventually collapsing due to over-exploitation of their resources. Today, technological progress has expanded our carrying capacity, allowing economies to become more productive and per capita incomes to grow rapidly. But, whilst the long-term trend of Western capitalism has seen economies grow and societies shift towards education and specialisation, it has been at the expense of our environment. The unprecedented growth of the world economy is clearly contributing to global warming. In 1992, Grossman and Krueger (1991) developed the Environmental Kuznets Curve (EKC), which proposed that initial increases in gross domestic poroduct (GDP) per head lead to increased GHG emissions but, as an economy transitions from industrial production to service-based industries, the environmental damage gradually falls (Figure 2.1). This model has now been validated both for emerging economies, where we continue to see GHG emissions rising, and more service-based economies, such as in the UK, where there is some evidence of GDP growth decoupling from environmental damage.

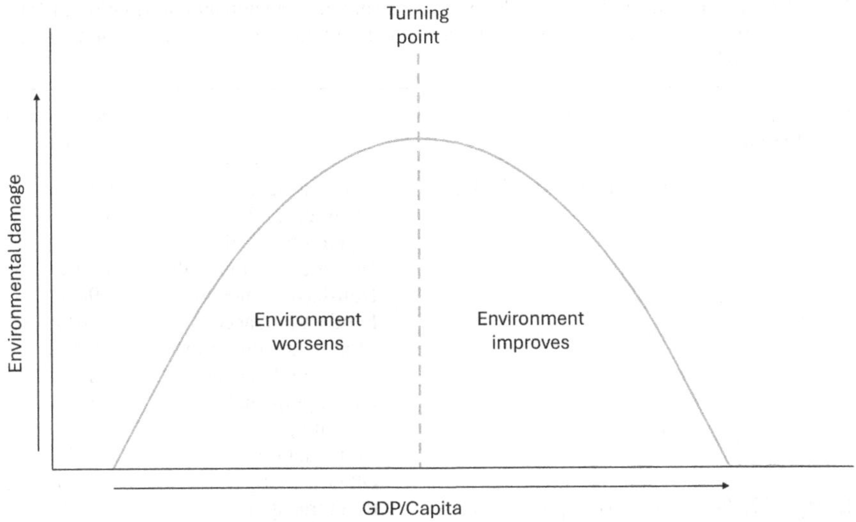

Figure 2.1 Environmental Kuznets curve, showing the relationship between greenhouse gas emissions and economic growth and potential for decoupling. Adapted from Grossman and Krueger (1991).

This shows that the carbon-intensive nature of the manufacturing sector and extractivism (the large-scale extraction of natural resources) are fundamental to climate change. This is important for understanding the role organisations can play in helping create a more sustainable future. Table 2.1 outlines the main sectors that are contributing to GHG emissions and shows their percentage of total GHG emissions, clearly demonstrating energy use as the largest contributor, foremost in industry and secondarily in transport.

Since 1990, the four emission sources that have grown the fastest are: industrial processes, which grew by a massive 225%; transportation (up 66%); electricity and heating (up 88%); and manufacturing and construction (up 60%). These should take priority in emission reduction strategies across all sectors of society, both at the individual and organisational levels. For example, do offices need to be heated throughout spring and autumn? Are workers driving to the office when they could cycle? Every action makes a difference. Current national policies (see Chapter 3) are on target to achieve less than a 1% reduction in GHG emissions by 2030. To have just a 50% chance of limiting global warming to 2°C, global emissions need to be cut by 28% by 2030. Rapid transformations are therefore needed across all sectors and systems. Organisations need to phase out coal in electricity generation, halt deforestation, increase the share of low-carbon fuels in transportation, and scale up public and private climate finance, among other actions.

Figure 2.2 illustrates the underlying organisational and consumer behaviours that are driving the current climate crisis, and demonstrates the feedback mechanisms, firstly between the human environmental systems, and secondarily between

Table 2.1 Sector contributions to GHG emissions showing the annual contribution of total anthropogenic emissions. Data derived from Climate Watch (2024) and Ritchie (2020).

Sector type (% emissions)	*Use in (% emissions)*	*Sector*	*Annual GHG emissions*
Energy (72.5%)	Industry (24.4%)	Iron and steel	6.10%
		Chemical and petrochemical	3.80%
		Non-metallic minerals	3.10%
		Non-ferrous metals	1.90%
		Food and tobacco	1.20%
		Mining and quarrying	0.80%
		Textile and leather	0.60%
		Paper, pulp, and printing	0.60%
		Construction	0.30%
		Other industry	6%
	Transport (13.6%)	Road transport	12.10%
		Aviation	0.70%
		Shipping	0.30%
		Rail	0.20%
		Pipeline	0.30%
	Buildings (18.9%)	Residential buildings	12.50%
		Commercial buildings	6.40%
	Other (15%)	Unallocated fuel combustion	8%
		Fugitive emissions from production	6.60%
		Energy in agriculture and fishing	1%
Industry (6.2%)		Cement	3.40%
		Chemicals	2.60%
		Electronics	0.10%
		Electric power systems	0.10%
Waste (3%)		Landfills	2%
		Wastewater	1.30%
Agriculture, Forestry, and Land Use (18.2%)		Livestock and manure	5.80%
		Agricultural soils	4.10%
		Crop burning	3.50%
		Deforestation	2.20%
		Cropland	1.40%
		Rice cultivation	1.20%

the organisational/individual behaviours that both drive harmful climate impacts as well as those helping to abate them. Understanding the contribution of specific sectors, as well as these feedback loops, is critical for knowing which actions and strategies we should be focusing on to catalyse the scale of change needed to avert the worst-case scenario outlined in Chapter 1. Figure 2.2 also highlights the way in

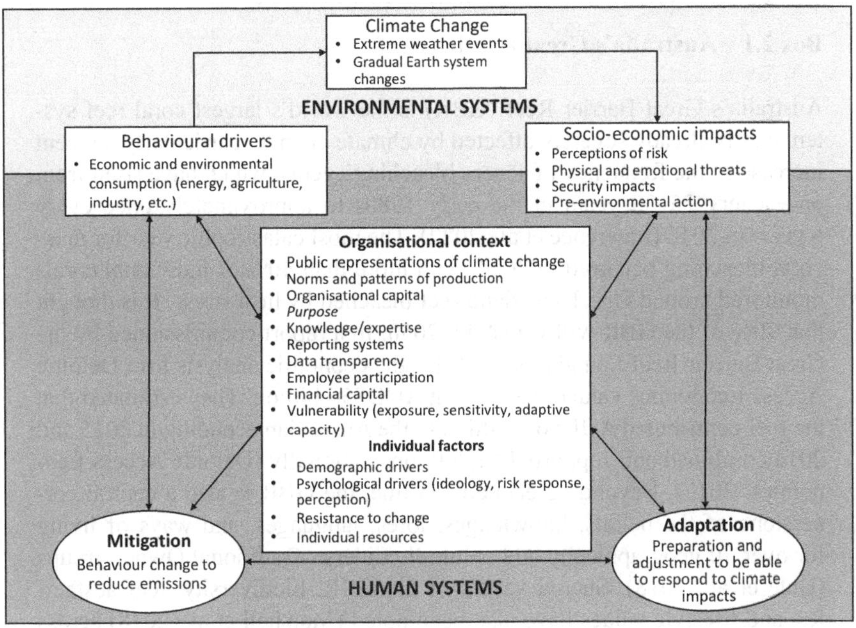

Figure 2.2 Behavioural drivers and feedback loops between climate change, socio-economic impacts, and organisational/individual action. Adapted from Gifford et al. (2011).

which organisations and individuals are at the heart of both the climate problem and its solution. In the remainder of this chapter, we will focus on the socio-economic impacts of climate change and demonstrate why there is not only a moral reason to mitigate the harmful effects of global warming but how there is also a logical and rational business case to be made for organisational climate action.

Climate change impacts on ecosystems and natural resources

The natural environment underpins our ability to function. Healthy land and ecosystems are essential for human health, food, and livelihood security (IPCC, 2019). Climate change, however, is accelerating a catastrophic loss of biodiversity and degradation of natural resources, posing significant threats to ecosystems and human well-being, which are expected to have cascading and long-term impacts on people globally (Pörtner et al., 2022). One of the primary drivers of biodiversity loss is habitat destruction caused by climate-related factors, such as wildfires, droughts, rising temperatures, and sea levels. Coral reefs, for example, are experiencing widespread bleaching due to increased ocean temperatures and acidification, leading to a decline in marine biodiversity which negatively impacts fisheries, as well as tourism that depend on these ecosystems (see Box 2.1).

Box 2.1 Australia's Great Barrier Reef

Australia's Great Barrier Reef (GBR) is the world's largest coral reef system and is already severely affected by climate change, particularly a recent increase in the frequency of severe bleaching events, which have risen from once every 25–30 years in the early 1980s to approximately once every 6 years in 2016 (Lawrence et al., 2022). The most catastrophic year for mass coral bleaching occurred in 2024, with more than 40% of individual corals monitored around One Tree Island reef bleached by heat stress. It is thought that 90% of the GBR will vanish by 2050. In a report commissioned by the Great Barrier Reef Foundation in 2017, the economic analysis firm Deloitte Access Economics valued the GBR at AUD$56 billion. They estimated that the reef contributed AUD$6.4 billion to the Australian economy in 2015 and 2016 combined and supported 64,000 jobs nationally (Deloitte Access Economics, 2017). Beyond its economic value, the GBR is also a central cornerstone of the beliefs, knowledges, lores, languages, and ways of living for over 70 geographically and culturally diverse Traditional Owner groups (Dale et al., 2018). Shared values of the GBR, biodiversity, reef aesthetics, and lifestyle values have also been noted (Marshall et al., 2018) across widely diverse stakeholder groups, including Australians living outside the catchments of the GBR, coastal communities, domestic and international tourists, and commercial fishers. This emphasises not only the economic and environmental value of ecosystems such as the GBR but also the cultural values. The implications of its loss as a source of revenue and employment for tourism, as well as the incommensurate value to indigenous people and broader national and international citizens, are enormous.

Beyond the tangible impacts associated with the loss of tourism-based revenue and employment, biodiversity loss also contributes to deeper, more systemic problems that affect entire ecosystems. Habitat loss and changes in seasonal climates are affecting species distributions and migration patterns. Many species are shifting their ranges towards higher latitudes or altitudes to escape rising temperatures, which can disrupt ecological interactions and food webs. Pollinators such as bees and butterflies, for example, are essential for global food production, pollinating an estimated 35% of global crop volume (IPBES, 2016) and are particularly vulnerable to climate shifts. Various pollinator species are also exhibiting range loss, reducing the number of pollinators for some crops. The buff-tailed bumblebee (*Bombus terrestris*), for instance, is a common visitor to many temperate crops and wildflowers and is projected to lose range in Southern Europe by 2100, resulting in an even greater decrease in pollination (Rasmont et al., 2015).

The following list indicates key ecosystem services (the benefits that humans derive from ecosystems) that are influenced by climate change and the effects this has on the ecosystem services they provide. We next outline some of these ecosystem services and their implications for businesses today:

- Reduced freshwater availability due to prolonged droughts and declining snowpack levels
  - Directly impacts agricultural productivity, leading to lower yields and food insecurity
  - Water is essential for industrial processes, and shortages can disrupt manufacturing and affect the overall economy
  - Access to clean water and sanitation is crucial for public health, and waterborne diseases are a major concern in water-scarce regions
  - Urban areas often face high water demand, and water scarcity can exacerbate social inequalities and create challenges for urban development
  - Water scarcity can lead to conflict and tension over limited resources
- Soil degradation, driven by increased temperatures and extreme weather
  - Increases soil erosion and reduces agricultural yields
  - Degraded soil contains fewer nutrients, impacting dietary nutrition and human health
  - As yields decrease, the cost of food increases, affecting consumers and exacerbating food insecurity
  - Soil degradation releases GHGs into the atmosphere, further exacerbating climate change
  - Degraded soil can lead to increased run-off, pollution, and reduced water infiltration, impacting water resources
- Ocean acidification and warming
  - Decline in the availability, distribution, and quality of commercially important species
  - Decreasing oxygen levels reduce average fish body size by 14–24%
  - Geographic range shifts for marine fish, arthropods, and marine mammals

Box 2.2 Organisational actions to avert biodiversity loss

- Invest in protected areas, including national parks and reserves
- Rewild around offices, including hedgerows and wildflower lawns
- Select industrial parks that are built with nature-positive principles, for example, integrated forests and green spaces
- Buy from sustainable producers, including regenerative farmers, FSC-certified foresters, and MSC-certified fisheries
- Reduce greenhouse gas emissions in operations
- Invest in conservation efforts
- Reduce consumption
- Plant native species and pollinators on business grounds, install bat boxes and bug hotels
- Minimise the use of harmful chemicals
- Raise awareness about biodiversity loss

Climate change and food security

A changing and more volatile climate is directly affecting the yields of food and fibre production globally. Floods in Bangladesh in 2024, for example, destroyed over 1 million tonnes of rice and 200,000 tonnes of vegetables. In Europe, the severity of heatwave and drought impacts on crop production roughly tripled over the last 50 years (Brás et al., 2021), whilst variability in yields has also increased significantly, making it harder for budgeting, cash flow, and value chains. More gradual changes, including seasonality, are beginning to alter the pattern and distribution of food production, and these types of changes are expected to continue and intensify over time. Warming temperatures, for example, can cause plants to flower and fruit earlier, potentially leading to mismatches between crop cycles and traditional harvest times. This can create challenges for post-harvest handling and storage, disrupting the flow of food along the supply chain and leading to shortages or surpluses and changing costs for both producers and consumers.

These changing yields are already affecting the price of several globally beloved products, including coffee, chocolate, and olive oil (Angel & Adombila, 2024). New York cocoa prices have been around US$2,000–US$3,000 a tonne for decades, yet had reached US$12,500 at the end of 2024. The lower supply and higher price in turn put pressure on processing companies, retailers, hospitality businesses, and ultimately consumers. Despite the massive rise in US cocoa prices, however, it is important to note that only 10% of the increase has been passed on to cocoa farmers in West Africa. Considering the climate risks that the cocoa industry is facing, such commoditisation, without producer investment, intensifies the vulnerabilities of the communities whose livelihoods depend on cocoa production.

Transportation and distribution networks are also vulnerable, as floods, storms, and rising sea levels can damage roads, railways, and ports, causing delays and increased costs. Processing and packaging industries face higher energy and water demands, with disruptions in raw material supply affecting production efficiency with increased prices again being passed on to retailers and consumers. Such fluctuations in both food price and availability are particularly harmful in regions that are highly dependent on imports. Vulnerable groups, such as women, children, low-income households, or other minority groups, are often at higher risk of the second-order impacts that can result from such fluctuations, including malnutrition, livelihood loss, rising costs, and competition over resources. These types of cascading risks emphasise how essential it is to invest in adaptive strategies such as climate-resilient infrastructure, improved storage technologies, and sustainable agricultural practices to ensure long-term food security.

Box 2.3 Are we witnessing the death of Champagne?

Up to 70% of the world's wine-producing regions could become unsuitable for growing grapes if global warming exceeds 2°C. Over the past 20 years, temperatures in the Champagne region have risen 1.2°C, accelerating bud-burst and bringing the harvest forward by an average of 18 days. These warmer summers generally mean faster-growing, sweeter grapes, creating Champagnes with ever-lower dosage or added sugar. For the big Champagne houses, which rely on preserving a distinctive house style year after year, these changes present increasingly difficult challenges. The rising unpredictability and extremity of the weather are also forcing vineyards to adapt their management strategies, making it hard to plan ahead for the growing season.

But, whilst the Champagne region struggles to adapt, Southern England is experiencing warmer and sunnier conditions, resulting in a climate that is very similar to that of Champagne 50 years ago. Now with over 500 vineyards in Sussex alone, a new generation of British winemakers is getting international recognition for their sparkling wines. This growth of English winemaking is having a transformative impact on traditional agriculture, altering the landscape and creating new economic opportunities for agricultural communities.

In some areas, there may also be opportunities to take advantage of the changes in climate and grow different crops compared with what was previously possible. Sorghum, for example, which is an arid-adapted crop, has shown consistent increases in global yields over the last 40 years, rising from c. 1.2 million tons per hectare in 1961 to c. 2.8 million tons per hectare in 2019 (Khalifa & Eltahir, 2023). However, the changes in extremes and variability of weather are likely to limit the magnitude of these types of opportunities. This changing suitability has important implications for regions that have built their reputation and infrastructure over many years around the production of particular crops and attracting a premium for that (see Box 2.3).

Box 2.4 Organisational actions to mitigate food security impacts

- Invest in technologies and agricultural practices that enhance productivity and resilience, for example, drought-resistant crops, efficient irrigation, and reduced tillage
- Diversify business catering, for example, buying regenerative farm products, and increase plant-based diets
- Develop policies to manage water demand, particularly in water-scarce regions
- Minimise food waste
- Support local food systems that reduce transportation distances and promote waste reduction
- For those working in the agri-food industry:
    - Build infrastructure, such as food storage facilities, that can withstand extreme weather events
    - Ensure access to food supplies in regions facing climate-related disruptions through open and equitable international trade
    - Support climate policies that promote sustainable agriculture
    - Increase public awareness of the challenges of climate change on food security and the importance of sustainable food systems
    - Ensure equitable payment to farmers and food producers in climate-vulnerable regions

Climate change and human health

The impacts of climate change on both physical and mental human health can be profound. Communicable and non-communicable diseases, such as dengue fever, Lyme disease, and cholera, are increasing and their distributions are changing with higher temperatures and heavy rainfall events. After extreme flooding in 2022, Pakistan experienced major outbreaks of cholera, malaria, and dysentery affecting hundreds of thousands of individuals (Ahmed & Ebrahim, 2022). Some climate-sensitive respiratory and allergic respiratory diseases are also changing, and climate-sensitive food-borne, waterborne, and vector-borne diseases are expected to increase in the coming decades (Cissé et al., 2022).

Box 2.5 Heat-related impacts on productivity

A recent study from the OECD (Costa et al., 2024) highlights the impact that heat stress can generate on the productivity of manufacturing and services firms. The authors used a data set of detailed weather and financial information for more than 2.7 million manufacturing and services firms across 23 advanced economies between 2000 and 2021, complemented with country-level information on adaptation investment. They found that both more frequent 'hot' days and the occurrence of heatwaves lead to substantially reduced labour productivity. Ten extra days above a temperature of 35°C in a year resulted in a 0.3% reduction in firms' annual labour productivity. One additional heatwave with temperatures above the 95th percentile of historic averages and lasting at least five days caused a reduction of up to 0.2% in firms' annual labour productivity. Smaller and less productive firms were found to be even more vulnerable to the impact of heat stress. A study on the credit lines issued by private banks to firms in Mexico (Aguilar-Gomez et al., 2024) found that extreme heat led to an increase in credit delinquency rates, which was primarily driven by small firms defaulting on their loans. The variation in impact across firms emphasises the difference in adaptive capacity between size of firms. Larger firms, for example, may be more resilient due to fewer financial constraints, better access to advanced technology, and knowledge of behavioural adaptation practices.

Extreme heat not only has negative physical effects (in extreme cases resulting in death) but has also been shown to be detrimental to mental health, well-being, life satisfaction, happiness, cognitive performance, and aggression (Thompson et al., 2023; Doherty & Clayton, 2011). A study from Australia found that violence, and particularly domestic violence, increased with extreme heat (Stevens et al., 2023). And in the case of the labour force, the combined physical and mental impacts of extreme heat can have significant effects on worker productivity, with some regions globally reaching the upper limits of labour productivity (see Box 2.5). Increasing global temperatures are being felt most acutely in urban environments, where the combination of artificial surfaces like buildings and roads, which absorb and retain more heat than natural surfaces, energy consumption in buildings, traffic and industrial process, which generate heat, and the reduced cooling effects of vegetation and waterways, leads to a phenomenon known as the Urban Heat Island effect (UHI). The UHI leads to many cascading impacts, such as increased demand for air conditioning and economic losses due to reduced work capacity, and increased energy costs. But perhaps the most tangible effect is on human health, leading to heatstroke, heat exhaustion, and other heat-related illnesses, particularly affecting vulnerable populations like the elderly and those with pre-existing conditions. The UHI can also worsen air quality, increasing exposure to pollutants, which can exacerbate respiratory and cardiovascular problems.

Box 2.6 Organisational actions to mitigate climate change impacts on human health

- Improve air quality by reducing fossil fuel consumption and transitioning to clean energy sources like solar and wind power
- Integrate green infrastructure into buildings to improve air quality, provide green spaces, insulate buildings, and reduce the urban heat island effect
- Encourage walking and cycling through investments in pedestrian and cycling infrastructure and employee support schemes
- Ensure healthcare system preparedness by investment/engagement with local healthcare providers
- Provide hydration and breaks to employees
- Ensure adequate ventilation and cool environments for employees to avoid heat stress
- Relax dress codes or introduce flexible hours or early/late starts to avoid high or low temperatures

Climate-driven resource conflicts and migration

Climate change is intensifying competition over scarce resources, leading to conflicts that disrupt business operations and threaten investment security, particularly in regions where governance is weak or where resources are already limited.

Water shortages

In many regions, water scarcity is exacerbating existing vulnerabilities, leading to social unrest and heightened risks for companies reliant on stable supply chains (Karamidehkordi et al., 2024). In the central Highlands of Mexico, for example, San Cristóbal, a city of 200,000 people, is home to a bottling plant run by Femsa, a food and drink conglomerate with rights to bottle and sell Coca-Cola in much of Latin America. It has permits to extract more than 300,000 gallons of water every day from the nearby Huitepec volcano basin. Yet, despite the region having the most water per capita in all of Mexico, more than one in three people in rural areas do not have access to running water. Mexico is projected to face water shortages in 20 of its 32 states by the year 2050 because of climate change and severe drought. As a result, conflicts over water rights have flared between locals and companies like Coca-Cola and the French multinational Danone (Yeung, 2025). Such pressures on business operations and local communities raise glaring questions that go beyond the mere physical insufficiency of resources, to highlight inadequacies in the social relations responsible for the allocation, use, and conservation of critical ecosystem services.

Migration

In recent years, war-torn regions such as Darfur in northern Sudan and Yemen have harboured severe humanitarian crises, both driven by and compounded by climate-related hazards. Large-scale migration is increasing as communities flee extreme weather events, desertification, conflict and violence, and rising sea levels. Whilst the displacement of people is not only attributable to climatic change, the number of people living in internal displacement has doubled since 2018, reaching a record high of 83.4 million across 117 countries and territories at the end of 2024 (IDMC, 2025). Future projections of climate-driven migration suggest that these numbers will only increase. The International Displacement Monitoring Centre, for example, predicts that 1.2 billion people could be displaced globally by 2050 due to a rise in extreme weather and natural disasters.

Labour markets

Climate-driven displacement is reshaping labour markets, presenting both challenges and opportunities for businesses and economies worldwide. In some regions, the influx of climate migrants can create labour surpluses, increasing competition for jobs and potentially lowering wages. In other areas, out-migration leads to labour shortages, disrupting industries such as agriculture, construction, and manufacturing. Countries and businesses that can effectively integrate migrant workers will benefit from new skills and labour force diversity, while those that fail to adapt may experience economic instability. Corporations can play a role in facilitating economic integration by supporting policies that promote fair labour conditions and workforce training for displaced populations. Companies that invest in adaptive hiring practices and sustainable job creation can contribute to economic stability while addressing workforce challenges. Similarly, comprehensive and consistent national policy is needed to support growing climate migrant populations. Whilst some countries have explicit climate policies that acknowledge a future rise in migration, there are currently no internationally recognised mechanisms to implement cross-border resettlement of climate migrants.

Box 2.7 Organisational actions to mitigate climate change-induced conflict

- Ensure operations do not negatively impact access to resources by local residents, or extract resources in unsustainable quantities, for example, water and minerals
- Work with intergovernmental and humanitarian organisations to strengthen climate action in conflict zones and to ensure that vulnerable populations are protected
- Proactively assess potential climate risks in business decision-making
- Develop policy to effectively integrate migrant workers to ensure consistency in labour supply
- Exploit reuse opportunities from waste to save money and lower your carbon footprint, for example, waste heat or water
- Promote fair labour conditions and workforce training to accommodate changing labour markets
- Integrate Climate Justice principles at all stages of business decision-making, for example, fair pay to Global South suppliers, and engagement of SMEs in supply chains
- Take a people-centred approach to environmental management

Impacts on communities at different scales

It is essential to recognise that the effects of climate change are not felt equally across all communities. Warmer temperatures, more extreme weather events, and disruptions to global supply chains disproportionately affect populations at different scales. This recognition of differential impacts on communities is at the heart of the concept of Climate Justice, a term that refers to the unequal impacts of climate change and the unequal burden of mitigation efforts on vulnerable populations.

At the macro-scale, the Global South experiences considerably more loss and damage than the Global North, despite the vast majority of historical emissions being attributed to the US (29%), EU (22%), and Russia (6%) (see Chapter 1). Ratings firm S&P Global published a recent report on the likely impact of rising sea levels, heatwaves, droughts, and storms attributable to anthropogenic climate change. Under a baseline scenario of limited climate action, they estimated that over 12% of GDP in low/middle-income countries was at risk compared to just 3% in upper-middle/upper-income countries (Jones, 2022). Bangladesh, India, Pakistan, and Sri Lanka's exposure to wildfires, floods, major storms, and water shortages mean that South Asia has 10%–18% of GDP at risk, roughly treble that of North America and ten times more than the least-affected region, Europe. What is clear is that climate change is disproportionately impacting the poorest communities (see Chapter 3 on the concept of *Common but Differentiated Responsibility*).

At more local scales, we see that this trend continues, with low-income populations, indigenous groups, and marginalised women and children, for example, facing heightened risks from climate-related disasters, food insecurity, and health crises. Marginalised communities often live in high-risk areas such as flood-prone coastal regions, arid zones susceptible to drought, or urban slums with poor drainage and infrastructure. When extreme events occur, these populations suffer the most due to inadequate housing, lack of insurance, and minimal access to emergency response services. For example, Hurricane Katrina in 2005 disproportionately affected low-income African American communities in New Orleans, highlighting the intersection of environmental and social injustices (Laksa & Morrow, 2006). Limited access to resources, infrastructure, and political representation further compounds their vulnerability, making adaptation and recovery significantly more challenging.

Climate change and infrastructure: Threats to business assets, operations, and supply chains

Climate change poses significant risks to corporate infrastructure, logistics hubs, and transportation networks. Rising sea levels, hurricanes, floods, wildfires, and extreme heatwaves are already causing direct damage to physical assets, leading to economic losses and operational disruptions (see Table 2.2 for the types of impacts on infrastructure).

Hurricane Harvey in 2017 led to substantial disruptions in Houston's shipping operations, affecting supply chains across multiple industries (Bathgate et al., 2022). The Port of Houston had to close for a week following the storm; flooding and subsequent silt hindered vessel traffic which, in turn, increased shipping costs and caused logistical disruptions for businesses throughout the supply chain. As key trade hubs, ports generate employment and connect national and international markets. This global integration means that even temporary port disruptions—such as Hurricane Harvey—can trigger widespread logistical bottlenecks. Approximately one-third of the world's ports are in areas exposed to tropical storms (Becker et al., 2018) but, despite this, very few port authorities have begun implementing adaptation strategies to boost their resilience.

Similarly, floods and landslides triggered by extreme rainfall can damage road and rail infrastructure. Intense rainfall in the Auckland region in New Zealand in 2023 led to flash flooding and widespread landsliding, including two fatal landslide events (Brook & Nicoll, 2024). The total cost of the storms exceeded NZ$14.5 billion (Pullar-Strecker, 2023), which was 3.4% of New Zealand's projected GDP for 2023 (Brook & Nicoll, 2024). Although extreme weather events are not new, they are expected to become more frequent and intense in a changing climate, and are already exacerbating existing vulnerabilities across corporate, logistical, and transportation infrastructures (see Table 2.2).

Similarly, extreme weather events are increasingly disrupting manufacturing processes and global trade. Direct impacts on sectors such as food, mining, and logistics cascade into other interlinked global supply chain networks (Ghadge et al., 2019). Flooding, for example, can submerge factories, damage equipment, and halt production for extended periods. The 2011 Thailand floods, which affected major industrial zones, led to significant disruptions in the global electronics and automotive industries (Haraguchi & Lall, 2015). Hurricane Harvey also caused

Table 2.2 Summary of potential climate impacts on infrastructure. Adapted from Dawson (2015).

	Climate impact																
	Sea level rise			*Rainfall*				*Temperature*					*Other impacts*				
	Coastal flooding	*Saline intrusion*	*Coastal erosion*	*Freshwater flooding*	*Drought*	*Biological processes/Disease*	*Landslides*	*Severe heat*	*Severe cold*	*Subsidence/ desiccation*	*Biological process/disease*	*Demand for service*	*Lightning*	*Humidity*	*Solar radiation*	*Fog*	*Sorm/wind damage*
Potable water supplies	X	X	X	X	X	X		X	X	X	X	X					X
Wastewater and storm water systems	X	X	X	X	X	X		X	X	X	X						X
Energy demand	X			X	X			X	X			X			X		X
Power systems and communication	X			X				X					X				X
Renewable energy generation	X			X	X										X		X
Ports and associated infrastructure	X		X	X				X	X							X	X
Rail transport	X		X	X	X		X	X	X	X		X	X	X	X		X
Road transport	X		X	X			X	X	X	X		X		X		X	X
Buildings	X		X	X			X	X	X	X		X	X	X			X
Airports	X		X	X	X		X	X	X	X		X	X			X	X
Information and technology	X			X	X			X	X				X	X			X

extensive damage to petrochemical plants along the U.S. Gulf Coast, leading to a ripple effect across industries reliant on these chemicals (Misuri et al., 2019). These disruptions underscore the need for businesses to reassess their manufacturing and supply chain strategies in light of increasing climate risks.

Ironically, climate impacts on energy infrastructure could create challenges for shifting away from fossil fuels. For example, droughts and heatwaves in 2021 and 2022 in China resulted in shutdowns to their hydropower facilities causing severe blackouts and creating political pressure for a continued expansion of coal, undermining the country's ability to reach its climate change emissions targets (Dineen, 2025).

Financial risks, disclosure, and insurance in the context of climate change

An increasing likelihood of climate-related extreme weather events is driving change in the availability and affordability of insurance (Pastor-Paz et al., 2020; Storey et al., 2024). Insurers are increasingly factoring in climate risks when assessing premiums, making it more expensive for homeowners and businesses to secure adequate coverage for their properties and operations. For instance, in California, wildfires led insurers to withdraw from certain markets, forcing businesses to seek coverage from government-backed programmes or self-insure at a higher cost (Isidore, 2025). Ultimately, it is possible that some areas will be unable to obtain insurance. Some countries and regions are exploring 'managed retreat' options, relocating assets in instances where the risk is very high. Alternatively, companies can invest in asset protection measures to adapt to the changing risk. Examples of this include retrofitting buildings with resilient materials, elevating structures in flood-prone areas, and enhancing drainage systems.

With the recognition that climate change will create risks and opportunities for the global financial sector, the Task Force on Climate-Related Financial Disclosures (TCFD) was formed by the Financial Stability Board in December 2015. The TCFD's remit was to provide recommendations for climate-related financial disclosures for firms in the financial sector. The TCFD identified two major risk categories—physical risks from climate change, and transitional risks (i.e. those involved in transitioning to a lower-carbon economy). Having fulfilled its remit in 2023 with two inaugural standards issued by the International Sustainability Board (ISSB), the TCFD disbanded, with monitoring being continued by the ISSB.

The creation of the TCFD raised global awareness about the fact that investors are exposed to climate-related risks and must be made aware of them so they can choose where to invest based on complete information. In 2024, 30 jurisdictions are using or are taking steps to introduce ISSB Standards in their legal or regulatory frameworks, representing 57% of global GDP; 40% of global market capitalisation; and more than half of global GHG emissions (IFRS, 2024). We discuss these financial issues in more detail in Chapter 10.

Box 2.8 Organisational actions to mitigate climate change impacts on infrastructure and financial risk

- Integrate climate risks into infrastructure planning and project selection
- Identify infrastructure systems and supply chain components that are most vulnerable to climate risks
- Retrofit existing infrastructure to be more resilient to changed weather conditions
- Add redundancy into infrastructure networks in order to provide viable alternatives when some parts of the network fail
- Build in flexibility so that infrastructure assets can be modified in future without incurring excessive cost
- Design systems that consider how changes in climate will alter supply, demand, and risks
- Incentivise reduced demand for services through behaviour change and the use of more efficient technologies
- Recognise that different infrastructure systems (e.g. water, energy, and transportation) are interconnected and that climate change can create cascading effects
- Use reporting and disclosure regimes and explore various funding sources to incentivise private investment in climate-resilient infrastructure
- Develop mechanisms for sharing information on climate risks, vulnerabilities, and adaptation strategies
- Invest in asset protection measures to adapt to changing climate risks
- Create detailed plans for adapting to the impacts of climate change, including measures to protect infrastructure and maintain essential services

Interconnected, cascading, cross-border risks

Most studies that examine the impacts of climate change take a sectoral approach, limiting analyses to the sector of interest. However, this narrow view overlooks the transmission of impacts across sectors and borders and fails to identify the comprehensive impact of climate change. Cascades result from interdependencies between systems and sub-systems of coupled natural and socio-economic systems in response to changes and feedback loops (Wunderling et al., 2023). A recent example of such cross-border cascading impacts is the food affordability crisis in Europe, resulting from inflation in food prices, caused by drought-induced crop failures in breadbasket regions such as Bulgaria, Spain, and Italy (Harvey, 2022).

It is also important to recognise that risks can arise from both impacts of and responses to climate change. An example of such a cascading response risk is in the UK's Environment Act target of increasing tree and woodland cover to 16.5% of England's land area by 2050. Whilst such government mandates open new investment opportunities for carbon offsetting and land acquisition, it poses significant risk of offshoring food production emissions (known as Carbon 'Leakage'). For example,

UK tree planting targets could result in agricultural land being taken out of production on UK soil, leading to increased imports of products such as beef and soy, which are driving deforestation rates in regions such as the Amazon rainforest. Another potential cascading impact of increasing tree cover may be through changing the structure of communities in the UK, with less labour required for forestry compared to agricultural or horticultural land, which can impact inflation, economic growth, and public finances. Therefore, recognising the systemic nature of risks, and their propensity to cascade through multiple system components, is essential to accurately assess and plan for the broader socio-economic impacts of climate change (Challinor et al., 2018).

Opportunities for innovation and green growth: How businesses can lead the transition

In the era of heightened consumer awareness and corporate accountability, businesses that neglect social responsibility face significant reputational risks. Climate change has intensified expectations for corporations to take proactive steps in environmental and social governance (ESG), and failure to do so can lead to public backlash, financial losses, and long-term brand damage. Over the last ten years, there have been numerous cases of corporate greenwashing, with organisations attempting to appear more sustainable than they really are, such as the case in 2019 when McDonald's introduced paper straws that turned out to be non-recyclable (Picheta, 2019). Or in 2022, when the Advertising Standards Authority banned adverts by HSBC that showcased their climate-friendly initiatives but failed to mention their considerable contribution to the climate crisis at the same time, which was estimated to be $8.7 billion of funding for new oil and gas exploitation in 2021 (Liang, 2022).

However, while climate change presents numerous challenges, it also offers businesses an opportunity to lead in climate justice and drive equitable solutions. Companies that prioritise sustainability, inclusivity, and resilience can create long-term value, contributing to social and environmental progress, while avoiding the pitfalls of greenwashing.

One key opportunity lies in sustainable innovation. Businesses can develop climate-friendly technologies, renewable energy solutions, and circular economy models that reduce environmental impact, promote economic inclusivity, and decrease inputs. Companies can support marginalised communities by sourcing from small-scale producers, investing in local climate adaptation projects, and ensuring fair labour conditions throughout their supply chains (see Chapter 4). By integrating social equity into their business models, corporations can foster economic empowerment while reducing vulnerability to climate change.

Investment in sustainable innovation, however, requires fundamentally engaging with Climate Justice. At the 28th session of the Conference of the Parties to the UN Framework Convention on Climate Change (COP28) in 2023, funding arrangements were operationalised to channel support for decarbonisation to countries and communities that are most vulnerable to the impacts of climate change (see Chapter 3). To keep this momentum going, businesses have an important role to play. The investor-led initiative Climate Action 100+ (CA100+) publishes annual updates on the climate performance of 150 of the world's largest corporate GHG emitters through its Net Zero Company Benchmark. The latest results, from 2024, indicate that corporate action on the Just Transition (a shift towards a sustainable economy that ensures fairness and well-being for all, especially those who may be negatively impacted by the transition) is still in its early stages, yet despite it being a year of unprecedented geopolitical shifts, institutional investors are pursuing a common-sense approach to act on climate risks and opportunities. The world's largest corporate emitting companies are now seeing that climate risk is a financial risk, and if they do not seize the opportunities offered by a Just Transition, they will be left behind. These initiatives and their association with purpose-led organisational change are explored in greater detail in the later chapters of this book.

Collaboration between businesses, governments, and civil society organisations is also essential to driving systemic change. Public–private partnerships are crucial for scaling up both climate mitigation and climate adaptation projects, for example, in building climate-resilient infrastructure, sustainable agriculture, and community-led conservation efforts. One example of a promising alignment between free market mechanisms and government incentive is in biodiversity credit schemes. These involve buying and selling measurable units of biodiversity that are intended to compensate for the residual impact on biodiversity caused by projects. There are local and national schemes, as well as international ones, and more recently, efforts have been made to establish a more unified approach. The UK, in partnership with France, for example, is leading efforts to establish a global biodiversity credit market to facilitate the sharing of best practice on the governance mechanisms for credit funding, monitoring regimes to ensure biodiversity improvements, and the fair distribution of income to indigenous peoples and local communities.

Investment in nature-based solutions

Nature-based solutions (NbS) or Ecosystem-based adaptation (EbA) refer to actions that protect, sustainably manage, and restore natural and modified ecosystems, with the aim of creating benefits for both people and nature. They often generate multiple benefits concurrently, for example, through flood protection, biodiversity enhancement, carbon sequestration, and urban cooling. The European Commission also recognises NbS as a way to realise socially inclusive green growth (European Commission, 2015).

Examples of opportunities for businesses to invest in NbS include:

- **Building-related green spaces:** For example, green roofs, green facades, and building-integrated agriculture. These NbS provide benefits of water retention, cooling, food security, and biodiversity and are typically implemented at a decentralised level by building owners or entrepreneurs
- **Urban green spaces:** For example, parks, urban trees, and green spaces connected to grey infrastructure like playgrounds and street trees. These provide benefits of urban cooling, carbon sequestration, biodiversity, recreation, and well-being
- **Regenerative agriculture:** For example, minimal or no-till, maintaining cover crops, rotational grazing, and agroforestry. These provide carbon sequestration, biodiversity, flood protection, drought tolerance, health benefits, and market opportunities for premium pricing

- **Green-blue infrastructure:** For example, integrated green and blue spaces such as sustainable urban drainage systems, lakes, permeable surfaces, and wetlands; restoration of rivers and floodplains; creating vegetation buffers along rivers; and generating benefits related to flood management, biodiversity, carbon sequestration, and recreation

Although the benefits of NbS are wide-ranging and well-recognised, a key barrier to their more widespread implementation is a lack of finance. Toxopeus and Polzin (2021) review the academic literature on financing barriers for urban NbS. They find that barriers fall into two main clusters: a lack of coordination between public and private finances; and limitations in valuation methods that do not fully capture the wide range of benefits, prioritising the short over the long term. This is supported by other studies (Calliari et al., 2022). Current biodiversity credit schemes (see previous example), such as the UK's Biodiversity Net Gain market, are a new landscape where developers and land managers interact to ensure development projects have a positive impact on biodiversity. However, while many private actors (such as real estate developers) benefit from NbS, there is an expectation that they will be provided entirely through public funds. Overcoming these valuation and coordination barriers could enable greater private investment in NbS, generating benefits for both businesses as well as the wider community.

One emerging area for potential upscaling of NbS is in a resilience solution that combines risk transfer (e.g. insurance) with risk reduction (e.g. hazard mitigation). Coastal areas, for example, are particularly at risk from climate change impacts, given the high concentration of economic activity and the exposure to cyclones, tsunamis, sea level rise, and other coastal hazards. The traditional response to these risks has been through insurance and adaptation, typically treated as two separate planning and financing processes. Investment in NbS, however, offers a mutually beneficial opportunity for both climate resilience and economic growth. Reguero and colleagues (2020) propose a resilience insurance mechanism that addresses this trade-off. An investment in adaptation to the hazard at the beginning of the period reduces the underlying risk; the risk-reduction effect is monetised via reduced premiums during the treaty duration. A fraction of the initial investment is amortised at the end of the treaty, creating an incentive for the risk owner to invest in risk-reducing infrastructure while part of the risk is transferred through an insurance cover that ensures a payout in case of a catastrophic loss event. Whilst in the early phases of development, this type of financing mechanism offers a legitimate means of upscaling NbS within the current socio-economic system.

Furthermore, ecosystem protection contributes to economic stability by safeguarding essential natural resources that society and businesses depend on. Companies in sectors such as agriculture, forestry, and fisheries rely on healthy ecosystems for raw materials, water supply, and climate regulation, which in turn provide basic necessities for the wider population, and inputs for the business that rely on particular resources. For example, maintaining peatlands and wetlands helps secure clean water resources, reducing costs for industries that require significant water inputs, such as beverage production and manufacturing.

Businesses that prioritise ecosystem protection and sustainability also gain competitive advantages in a marketplace increasingly driven by environmental, social, and governance (ESG) criteria. Investors and consumers are demanding greater accountability from corporations regarding their environmental impact. Companies that integrate NbS into their operations can enhance brand reputation, attract investment, and improve stakeholder relations. In November 2024, for example, Triodos Bank published a White paper on 'Financing the Nature-based Solutions Sector' and announced its commitment to provide at least €500 million in investments, loans, and contributions to NbS projects by the end of 2030 as part of its biodiversity targets. At the heart of these targets is a committed purpose, described by Triodos Banks as a 'commitment (that) aligns with the bank's mission to help create a society that protects and promotes the quality of life of all its members, and that has human dignity at its core.'

Ultimately, the business case for ecosystem protection is clear: investing in NbS not only helps address climate change and biodiversity loss and other environmental and social concerns but can also strengthen long-term business resilience, enhance market opportunities, and contribute to a more sustainable global economy. By integrating nature into corporate strategies, businesses can play a pivotal role in advancing environmental sustainability while securing their financial future to generate new business opportunities, as we discuss further in the later chapters of the book.

Conclusion

Whilst the financialised socio-economic system of today may appear to some as being divorced from the natural world, the reality is that we rely on nature for our well-being and survival. The climate change impacts that we are already experiencing are only increasing as our emissions go unabated. Without significant upscaling of investment into renewable energy, green infrastructure, and nature-based solutions, we are likely to hit several catastrophic tipping points in the next 50 to 100 years. Almost all of the impacts will directly or indirectly affect businesses and organisations in some way. The most obvious impacts are the direct impacts from extreme weather events and supply chain disruptions. However, the indirect impacts on the systems that underpin so many parts of our economy—food systems, biodiversity, and water, for example—will eventually permeate through the system to impact our entire global economy. Climate change, however, is presenting new opportunities for investment into green technology and nature-positive action. To achieve the scale of uptake required to minimise the damage caused by historical and current emissions, organisations are going to need to adopt purpose-led organisational change. This chapter has attempted to illustrate the complex and systemic risks that businesses face if they do not rise to the challenge.

Further reading

Gates, M. 2021. *How to avoid a climate disaster. The solutions we have and the breakthroughs we need.* Knopf.

IPCC. (2022). Summary for policymakers. In H.-O. Pörtner, D. C. Roberts, E. S. Poloczanska, K. Mintenbeck, M. Tignor, A. Alegría, M. Craig, S. Langsdorf, S. Löschke, V. Möller, A. Okem (Eds.), *Climate change 2022: Impacts, adaptation, and vulnerability. Contribution of working group II to the sixth assessment report of the intergovernmental panel on climate change* (pp. 3–33). Cambridge University Press. https://doi.org/10.1017/9781009325844.001

Klein, N. (2014). *This changes everything: Capitalism vs. the climate*. Simon and Schuster.

References

Aguilar-Gomez, S., Gutierrez, E., Heres, D., Jaume, D., & Tobal, M. (2024). Thermal stress and financial distress: Extreme temperatures and firms' loan defaults in Mexico. *Journal of Development Economics*, *168*, 103246. https://doi.org/10.1016/j.jdeveco.2023.103246

Ahmed, K., & Ebrahim, Z. T. (2022). Health officials warn of major outbreaks of disease after severe floods in Pakistan. *The Guardian*. https://www.theguardian.com/global-development/2022/aug/31/health-officials-warn-of-major-outbreaks-of-disease-after-severe-floods-in-pakistan

Angel, M., & Adombila, M. A. (2024). Exclusive: African cocoa plants run out of beans as global chocolate crisis deepens. *Reuters*. https://www.reuters.com/markets/commodities/african-cocoa-plants-run-out-beans-global-chocolate-crisis-deepens-2024-03-13/

Bathgate, K., de la Cruz, A. P., & Zhang, Z. (2022). Quantitative analysis of hurricane Harvey impacts on Texas maritime facilities. *Transportation Research Record*, *2676*(7), 411–422. https://doi.org/10.1177/03611981221078574

Becker, A., Ng, A. K. Y., McEvoy, D., & Mullett, J. (2018). Implications of climate change for shipping: Ports and supply chains. *WIREs Climate Change*, *9*(2), e508. https://doi.org/10.1002/wcc.508

Brás, T. A., Seixas, J., Carvalhais, N., & Jägermeyr, J. (2021). Severity of drought and heat-wave crop losses tripled over the last five decades in Europe. *Environmental Research Letters*, *16*(6), 65012. https://doi.org/10.1088/1748-9326/abf004

Brook, M. S., & Nicoll, C. (2024). Brief report of fatal rainfall-triggered landslides from record-breaking 2023 storms in Auckland, New Zealand. *Landslides*, *21*(7), 1581–1589. https://doi.org/10.1007/s10346-024-02258-0

Calliari, E., Castellari, S., Davis, M., Linnerooth-Bayer, J., Martin, J., Mysiak, J., Pastor, T., Ramieri, E., Scolobig, A., Sterk, M., Veerkamp, C., Wendling, L., & Zandersen, M. (2022). Building climate resilience through nature-based solutions in Europe: A review of enabling knowledge, finance and governance frameworks. *Climate Risk Management*, *37*, Article 100450. https://doi.org/10.1016/j.crm.2022.100450

Challinor, A. J., Adger, W., Benton, T., Conway, D., Joshi, M., & Frame, D. (2018). Transmission of climate risks across sectors and borders. *Philosophical Transactions Royal Society, A*, *376*(2121).

Cissé, G., McLeman, R., Adams, H., Aldunce, P., Bowen, K., Campbell-Lendrum, D., Clayton, S., Ebi, K. L., Hess, J., Huang, C., Liu, Q., McGregor, G., Semenza, J., & Tirado, M. C. (2022). Health, wellbeing, and the changing structure of communities. In H.-O. Pörtner, D. C. Roberts, M. Tignor, E. S. Poloczanska, K. Mintenbeck, A. Alegría, M. Craig, S. Langsdorf, S. Löschke, V. Möller, A. Okem, & B. Rama (Eds.), *Climate change 2022: Impacts, adaptation and vulnerability. Contribution of working Group II to the sixth assessment report of the intergovernmental panel on climate change* (pp. 1041–1170). Cambridge University Press. https://doi.org/10.1017/9781009325844.009

Climate Watch. (2024). World Resources Institute.

Costa, H., Franco, G., Unsal, F., Mudigonda, S., & Caldas, M. P. (2024). *The heat is on: Heat stress, productivity and adaptation among firms* (No. 1828). OECD Publishing.

Dale, A., Wren, L., Fraser, D., Talbot, L., Hill, R., Evans-Illidge, L., Forester, T., Winer, M., George, M., Gooch, M., Hale, L., Morris, S., & Carmody, J. (2018). *Traditional owners of the great barrier Reef: The next generation of reef 2050 actions*. External Commissioned Report. Australian Government, Department of Agriculture, Water and the Environment.

Dawson, R. J. (Ed.). (2015). *A climate change report card for infrastructure*. LWEC Report Card. Living with Environmental Change. ISBN 978-0-9928679-4-2

Deloitte Access Economics. (2017). *At what price? The economic, social and icon value of the Great Barrier Reef*. https://www2.deloitte.com/au/en/pages/economics/articles/great-barrier-reef.html#

Dineen, J. (2025). Extreme weather could disrupt China's renewable energy boom. *New Scientist*. https://www.newscientist.com/article/2474942-extreme-weather-could-disrupt-chinas-renewable-energy-boom/

Doherty, T. J., & Clayton, S. (2011). The psychological impacts of global climate change. *American Psychologist*, *66*(4), 265–276. https://doi.org/10.1037/a0023141

European Commission. (2015). *Towards an EU research and innovation policy agenda for nature-based solutions & re-naturing cities*. Horizon 2020-European Commission [Internet]. https://ec.europa.eu/programmes/horizon2020/en/news/towards-eu-research-and-innovation-policy-agenda-nature-based-solutions-re-naturing-cities

Ghadge, D. A., Hendrik, W., & Seuring, S. (2019). Managing climate change risks in global supply chains: A review and research agenda. *International Journal of Production Research, 58*. https://doi.org/10.1080/00207543.2019.1629670

Gifford, R., Kormos, C., & McIntyre, A. (2011). Behavioral dimensions of climate change: Drivers, responses, barriers, and interventions. *Wiley Interdisciplinary Reviews: Climate Change*, *2*(6), 801–827. https://doi.org/10.1002/wcc.143

Grossman, G. M., & Krueger, A. B. (1991). *Environmental impacts of a North American free trade agreement*. National Bureau of Economic Research, Working Paper No. 3914. https://doi.org/10.3386/w3914

Haraguchi, M., & Lall, U. (2015). Flood risks and impacts: A case study of Thailand's floods in 2011 and research questions for supply chain decision making. *International Journal of Disaster Risk Reduction, 14*, 256–272. https://doi.org/10.1016/j.ijdrr.2014.09.005

Harvey, F. (2022). Falls in Europe's crop yields due to heatwaves could worsen price rises. *The Guardian*. https://www.theguardian.com/environment/2022/jul/27/big-falls-in-crop-yields-across-europe-feared-due-to-heatwaves

IDMC. (2025). *Global report on internal displacement 2025*. Internal Displacement Monitoring Centre.

IFRS. (2024). *Progress on corporate climate-related disclosures—2024 report*. https://www.ifrs.org/content/dam/ifrs/supporting-implementation/issb-standards/progress-climate-related-disclosures−2024.pdf

IPBES. (2016). Summary for policymakers of the methodological assessment report on scenarios and models of biodiversity and ecosystem services of the Intergovernmental Science-Policy Platform on Biodiversity and Ecosystem Services. In S. Ferrier, K. N. Ninan, P. Leadley, R. Alkemade, L. A. Acosta, H. R. Akçakaya, L. Brotons, W. W. L. Cheung, V. Christensen, K. A. Harhash, J. Kabubo-Mariara, C. Lundquist, M. Obersteiner, H. M. Pereira, G. Peterson, R. Pichs-Madruga, N. H. Ravindranath, C. Rondinini, & B. A. Wintle (Eds.). *Secretariat of the intergovernmental science-policy platform on biodiversity and ecosystem services*. IPBES.

IPCC. (2019). *Climate change and land: An IPCC special report on climate change, desertification, land degradation, sustainable land management, food security, and greenhouse gas fluxes in terrestrial ecosystems*. P. R. Shukla, J. Skea, E. Calvo Buendia, V. Masson-Delmotte, H.-O. Pörtner, D. C. Roberts, P. Zhai, R. Slade, S. Connors, R. van Diemen, M. Ferrat, E. Haughey, S. Luz, S. Neogi, M. Pathak, J. Petzold, J. Portugal Pereira, P. Vyas, E. Huntley, . . . J. Malley, (Eds.). Cambridge University Press. https://doi.org/10.1017/9781009157988

Isidore, C. (2025). California's insurance is in crisis. The solution will cost homeowners a ton. *CNN Business*. https://edition.cnn.com/2025/01/09/business/california-wildfires-homeowners-insurance

Jones, M. (2022). Climate change putting 4% of global GDP at risk, new study estimates. *Reuters*. https://www.reuters.com/world/climate-change-putting-4-global-gdp-risk-new-study-estimates-2022-04-27/

Karamidehkordi, E., Karimi, V., Singh, G., & Naderi, L. (2024). Water conflicts and sustainable development: Concepts, impacts, and management approaches. In S. A. Bandh & F. A. Malla (Eds.), *Water footprints and sustainable development* (Vol. 8, pp. 233–244). Elsevier. https://doi.org/10.1016/B978-0-443-23631-0.00016-9

Khalifa, M., & Eltahir, E. A. (2023). Assessment of global sorghum production, tolerance, and climate risk. *Frontiers in Sustainable Food Systems*, *7*, Article 1184373. https://doi.org/10.3389/fsufs.2023.1184373

Kotz, M., Levermann, A., & Wenz, L. (2024). The economic commitment of climate change. *Nature*, *628*(8008), 551–557. https://doi.org/10.1038/s41586-024-07219-0

Laksa, A., & Morrow, B.B (2006). Social vulnerabilities and hurricane Katrina: An unnatural disaster in New Orleans. *Marine Technology Society Journal*, *40*(4), 16–26.

Lawrence, J., Mackey, B., Chiew, F., Costello, M. J., Hennessy, K., Lansbury, N., Nidumolu, U. B., Pecl, G., Rickards, L., Tapper, N., & Woodward, A. W. (2022). Australasia. In S. L. H.-O. Pörtner, D. C. Roberts, M. Tignor, E. S. Poloczanska, K. Mintenbeck, A. Alegría, M. Craig, S. Langsdorf, B. R. V. Möller, & A. Okem (Eds.), *Climate change 2022: Impacts, adaptation and vulnerability. Change, contribution of Working Group II to the sixth assessment report of the intergovernmental panel on climate* (pp. 1581–1699). Cambridge University Press. https://doi.org/10.1017/9781009325844.013

Liang, A. (2022). HSBC climate change adverts banned by UK watchdog. *BBC News*. https://www.bbc.co.uk/news/business-63309878

Marshall, N., Barnes, M. L., Birtles, A., Brown, K., Cinner, J., Curnock, M., Eakin, H., Goldberg, J., Gooch, M., Kittinger, J., Marshall, P., Manuel-Navarrete, D., Pelling, M., Pert, P. L., Smit, B., & Tobin, R. (2018). Measuring what matters in the Great Barrier Reef. *Frontiers in Ecology and the Environment*, *16*(5), 271–277.

Misuri, A., Casson Moreno, V., Quddus, N., & Cozzani, V. (2019). Lessons learnt from the impact of hurricane Harvey on the chemical and process industry. *Reliability Engineering & System Safety*, *190*, 106521. https://doi.org/10.1016/j.ress.2019.106521

Newman, R., & Noy, I. (2023). The global costs of extreme weather that are attributable to climate change. *Nature Communications*, *14*(1), 6103. https://doi.org/10.1038/s41467-023-41888-1

Pastor-Paz, J., Noy, I., Sin, I., Sood, A., Fleming-Munoz, D., & Owen, S. (2020). Projecting the effect of climate change on residential property damages caused by extreme weather events. *Journal of Environmental Management*, *276*, Article 111012. https://doi.org/10.1016/j.jenvman.2020.111012

Picheta, R. (2019). McDonald's new paper straws aren't recyclable—but its axed plastic ones were. *CNN Business*. https://edition.cnn.com/2019/08/05/business/mcdonalds-paper-straws-recyclable-scli-gbr-intl/index.html

Pörtner, H.-O., Roberts, D. C., Poloczanska, E. S., Mintenbeck, K., Tignor, M., Alegría, A., Craig, M., Langsdorf, S., Löschke, S, Möller, V., & Okem, A. (2022). Summary for policymakers. In H.-O. Pörtner, D. C. Roberts, M. Tignor, E. S. Poloczanska, K. Mintenbeck, A. Alegría, M. Craig, S. Langsdorf, S. Löschke, V. Möller, A. Okem, & B. Rama (Eds.), *IPCC 2022: Climate change 2022: Impacts, adaptation and vulnerability. Contribution of working Group II to the sixth assessment report of the intergovernmental panel on climate change* (pp. 3–33). Cambridge University Press. https://doi.org/10.1017/9781009325844.001

Pullar-Strecker, T. (2023). *Repair bill from cyclone and Auckland floods at least $9b, Treasury estimates*. https://www.stuff.co.nz/business/131883544/repair-bill-from-cyclone-and-auckland-floods-at-least-9b-treasury-estimates

Rasmont, P., Franzen, M., Lecocq, T., Harpke, A., Roberts, S. P., Biesmeijer, J. C., Castro, L., Cederberg, B., Dvořák, L., Fitzpatrick, Ú., Gonseth, Y., Haubruge, E., Mahé, G., Manino, A., Michez, D., Neumayer, J., Ødegaard, F., Paukkunen, J., Pawlikowski, T., . . .

Schweiger, O. (2015). Climatic risk and distribution atlas of European bumblebees. *BIORISK ? Biodiversity and Ecosystem Risk Assessment*, *10*, 1–236. Pensoft Publishers.

Reguero, B. G., Beck, M. W., Schmid, D., Stadtmüller, D., Raepple, J., Schüssele, S., & Pfliegner, K. (2020). Financing coastal resilience by combining nature-based risk reduction with insurance. *Ecological Economics*, *169*, Article 106487. https://doi.org/10.1016/j.ecolecon.2019.106487

Ritchie, H. (2020). Sector by sector: Where do global greenhouse gas emissions come from? *Our World in Data*. https://ourworldindata.org/ghg-emissions-by-sector

Stevens, H. R., Graham, P. L., Beggs, P. J., & Hanigan, I. C. (2023). No retreat from the heat: Temperature-related risk of violent assault is increased by being inside. *International Journal of Urban Sciences*, *28*(1), 124–139. https://doi.org/10.1080/12265934.2023.2209544

Stern, N. (2006). *Stern review: The economics of climate change*. Cambridge University Press.

Storey, B., Owen, S., Zammit, C., & Noy, I. (2024). Insurance retreat in residential properties from future sea level rise in Aotearoa New Zealand. *Climatic Change*, *177*(3), 44. https://doi.org/10.1007/s10584-024-03699-1

Thompson, R., Lawrance, E. L., Roberts, L. F., Grailey, K., Ashrafian, H., Maheswaran, H., Toledano, M. B., & Darzi, A. (2023). Ambient temperature and mental health: A systematic review and meta-analysis. *The Lancet Planetary Health*, *7*(7), e580–e589. https://doi.org/10.1016/S2542-5196(23)00104-3

Toxopeus, H., & Polzin, F. (2021). Reviewing financing barriers and strategies for urban nature-based solutions. *Journal of Environmental Management*, *289*, Article 112371. https://doi.org/10.1016/j.jenvman.2021.112371

Wunderling, N., Winkelmann, R., Rockström, J., Loriani, S., Armstrong McKay, D. I., Ritchie, P. D., Sakschewski, B., & Donges, J. F. (2023). Global warming overshoots increase risks of climate tipping cascades in a network model. *Nature Climate Change*, *13*(1), 75–82. https://doi.org/10.1038/s41558-022-01545-9

Yeung, P. (2025). 'Forgotten'. How one Mexican city struggles against big industry for water. *Al Jazeera*. https://www.aljazeera.com/news/longform/2025/1/8/forgotten-how-one-mexican-city-struggles-against-big-industry-for-water

3 The global context for organisational climate action

Duncan Russel and Nick Kirsop-Taylor

Key points

- Climate action is based on a variety of global, national, and local policy-oriented organisations operating at different scales of governance
- The United Nations Framework Convention on Climate Change is the central global organisational actor coordinating and mediating between and across different political interests
- The explosion of policy-aligned organisations engaged with and around the UNFCCC has led to new and innovative ways to manage the burgeoning global network of actors

Introduction

Climate action is inherently multi-level and multi-organisational (Bulkeley, 2014). While the causes of the climate crisis are often (but not exclusively) located within states, the effects of climate breakdown are global, disaggregated, and misaligned with the human boundaries of the state (Lieven, 2021). Climate action is, therefore, the shared and common responsibility of all global communities and societies. In part, the global locus for climate action is anchored in the United Nations Framework Convention on Climate Change (UNFCCC) which convenes annual conferences of the parties (COPs) comprising nation-state representatives meeting to agree on a common approach to tackling global climate change.

In addition to nation states, there are other transnational organisations associated with the UNFCCC (Abbott, 2014; Hale, 2020; Keohane & Victor, 2011; Vogler, 2016) representing diverse groups ranging from city administrations through to business, indigenous groups and not-for-profit, environmental, and development organisations. These organisations coordinate at the level of the UNFCCC, as well as across national states and regions of the global, but also at national and sub-national levels of governance. This ecosystem of policy-facing organisations therefore performs a myriad of roles and functions, such as mediation and advocacy, delivering good governance and compliance outcomes (Biermann et al., 2009), securing climate finance from transnational capital (Singh & Jayaram, 2021), linking science to policy (Hoppe et al., 2013), and many others. These organisations

DOI: 10.4324/9781003545224-4

must be considered in terms of their inherent networked natures—their relationships with other organisations and actors of global climate governance.

In this chapter, we attempt to bring clarity to this complexity. First, we establish the nature of this complex modern organisational setting for governing climate change. Second, we outline the global climate governance context and the role of organisations therein. Third, we consider the national and sub-national government actors.

A brief history of international climate policymaking

Against the background of transnational and multi-level climate governance, the key international arena for global climate change action is the UNFCCC (Victor & Keohane, 2010), which entered into force on 21 March 1994. The work of the UNFCCC, its annual COPs, and associated working groups and subcommittees are administratively supported by a United Nations permanent transnational organisation/secretariat campus in Bonn, Germany. The UNFCCC COP brings together governments from around the world to discuss, bargain, and (hopefully) agree on measures to reduce greenhouse gas (GHG) emissions, adapt to climate impacts, and provide financial and technological support to vulnerable countries. The underlying objectives underpinning the UNFCCC process are to stabilise greenhouse gas emissions to prevent dangerous climate impacts, and to promote sustainable development and resilience with a focus on ensuring equality and the needs of developing and vulnerable countries. Critically, the central organisational decision-makers at the UNFCCC (manifest in its COP) are sovereign member states. These sovereign actors bring to the COP government officials, scientists, and national organisations of varying interests and constitutions as part of their negotiating teams. At the same time, fringe events to COP attract international businesses, environmental and development NGOs, as well as loose transnational networks from the third sector representing anything from indigenous groups, through to faith groups (also see the following text).

Since its inception in 1994, the UNFCCC has sought to coordinate a global approach to climate change based on the concept of common but differentiated responsibility (see Box 3.1 for its main actions and outcomes). The first meeting of the COP, which took place in Berlin in 1995, concluded that the commitments from the most developed countries (Annex I Parties) were not adequate. This led to the adoption of the Berlin Mandate, which provided a detailed process and timeline to negotiate more concrete obligations and time frames for the reduction of GHG emissions (https://legal.un.org/avl/ha/kpccc/kpccc.html). This eventually led to the first international agreement on climate change being signed in 1997, the so-called Kyoto Protocol.

Box 3.1 Major UNFCCC actions and agreements

1. **1992—UNFCCC adoption (Rio Earth Summit)**
   - Established the framework for international climate action
   - Signed by 154 countries; entered into force in 1994
2. **1997—Kyoto Protocol**
   - First legally binding treaty to reduce greenhouse gas emissions
   - Set binding targets for developed countries
   - Entered into force in 2005
3. **2009—Copenhagen Accord (COP15)**
   - Recognised the need to limit global warming to below 2°C
   - Introduced voluntary emission reduction pledges
4. **2015—Paris Agreement (COP21)**
   - Legally binding agreement to limit global temperature rise to well below 2°C (preferably 1.5°C)
   - All countries submit Nationally Determined Contributions (NDCs)
   - Entered into force in 2016
5. **2021—Glasgow Climate Pact (COP26)**
   - Strengthened commitments to phasing down coal and increasing climate finance
   - Urged countries to update their NDCs more frequently
6. **2023—COP28 (Dubai, UAE)**
   - The Global Stocktake (GST) was conducted to assess progress on the Paris Agreement goals
   - For the first time, an agreement mentioned the need for transitioning away from fossil fuels (although with vague commitments)
   - Countries reaffirmed the goal of tripling renewable energy capacity by 2030

The Kyoto Protocol set binding emission reduction targets for developed countries with the hope of reducing GHG emissions by 5.2%, and introduced instruments such as the Clean Development Mechanism, where organisations in Annex I countries, principally business, could offset their GHG emissions through investments in sustainable development projects in less developed countries. More developed Annex I countries agreed to reduce their emissions by 5.2% below 1990 levels during the first commitment period 2008–2012. Of note, some governments set differential targets, with, for example, EU member states agreeing to cut emissions by 8% overall but with individual targets within this envelope for different member states, such as the UK, which committed to a 12% GHG emissions reduction.

During the second commitment period agreed at the Doha COP in 2012, the most developed Annex I countries agreed to even more challenging emissions reduction targets of at least 18% below 1990 levels, but this was not widely ratified. Reporting and monitoring requirements were also prioritised with, for instance, Annex I countries having to report detailed emission inventories to the UNFCCC. Moreover, there was a commitment to support developing nations financially, through technology transfer and to build capacity to meet climate challenges.

While the Kyoto Protocol led to successful action in some instances, and provided a framework for the development and deployment of low-carbon technologies and innovative policy such as the European Union's Emissions Trading Scheme, overall global emissions actually rose 44% above 1990 levels by 2012 (including in many Annex I countries). In many respects, this outcome was not surprising given that many large growing economies with associated high and increasing GHG emissions, such as China, India, and Brazil, were not listed as Annex I countries and therefore had no requirement to reduce emissions. Moreover, the world's largest emitter at the time, the USA, failed to ratify the treaty despite the support of the then-President Bill Clinton. This was due to universal bipartisan opposition in the US Senate, fuelled largely by economic competitiveness concerns. In addition, only a few countries signed up to a new second commitment period (2012–2020) (e.g. the European Union, Australia, Norway, and Switzerland), with major economies such as Japan, Russia, Canada, and New Zealand not participating. Notably, Canada, for instance, withdrew from the Protocol in 2011 during COP15 because of this failure to reach a more comprehensive agreement. Indeed, COP15 was considered a failure more widely and provides a clear example of the challenges faced in achieving a comprehensive and meaningful international agreement on climate change—see Box 3.2 for details. Thus, the wider aim of reducing global GHG emissions was not realised.

Box 3.2 The failure of COP15

The United Nations Framework Convention on Climate Change (UNFCCC) COP15, held in Copenhagen in 2009, is widely considered a failure for several key reasons:

1. **Lack of a legally binding agreement**
   - The summit aimed to create a legally binding successor to the Kyoto Protocol but failed to do so
   - Instead, it resulted in the Copenhagen Accord, a non-binding agreement with vague commitments
2. **Weak emission reduction targets**
   - While the Accord acknowledged the need to limit global warming to below 2°C, it lacked specific and enforceable emission reduction commitments
   - Many developing nations, especially those most vulnerable to climate change, were disappointed by the lack of ambition
3. **Failure in negotiation process**
   - The negotiations were highly divisive, with major disagreements between developed and developing nations
   - Wealthy nations (like the USA and the EU) pushed for voluntary commitments, while developing nations (like China and India) demanded more responsibility from historically high-emission countries
   - The final deal was drafted in a closed-door meeting between the USA, China, India, Brazil, and South Africa, leaving out many countries, leading to accusations of an undemocratic process
4. **Lack of trust between nations**
   - Developing nations were frustrated that the promised $100 billion annual climate finance fund (to help them mitigate and adapt to climate change) had no clear implementation plan
   - Many developing countries felt sidelined, increasing tensions in future negotiations
5. **Public protests and discontent**
   - The summit saw massive protests from climate activists demanding stronger action
   - There were also large-scale arrests, adding to the negative perception of the event

While the Kyoto Protocol was only meant to be in place until 2012, disagreements between member states occurred over what a replacement treaty should look like, with tensions developing around whether emissions reduction targets should be placed on large emerging economies such as China, India, and Brazil whose contributions to GHG emissions were matching, and in some cases exceeding, those of developed countries, despite their significantly lower historical contribution to GHG levels.

Eventually, the Paris Agreement was signed in 2015 which replaced Kyoto targets and instead outlined commitments for all countries. In doing so, the overall aim of the agreement was to limit any global temperature rise to no more than 2°C, but preferably 1.5°C, to avoid the most dangerous impacts of climate change. Under the Paris Agreement, nation states defined their own targets (updated every five years), known as Nationally Determined Contributions (NDCs). The agreement also emphasised the importance of adaptation, technology transfer, and climate finance from richer to poorer nation states. On the issue of climate finance, developed countries had pledged $100 billion per year by 2020 to support developing nations, although this was never actually met—see Box 3.3 for examples.

Since the Paris Agreement was signed, UNFCCC negotiations have continued to evolve. As well as a focus on implementing the agreed NDCs, efforts have been made to make the emissions reduction targets of nation states more ambitious and to address perceived shortfalls in finance and equity in relation to those more vulnerable countries and communities. There are, though, a number of challenges that remain post-Paris—key among these is the difficulty of limiting warming to 1.5°C or below, and the fact that the emissions reductions committed to in NDCs do not match this ambition. Indeed, some scientific studies indicate that we may already be at the 1.5°C threshold (Bevacqua et al., 2025). Moreover, developing countries face delays and uncertainty in receiving promised funds from developed nations.

Negotiating dynamics: Why is a comprehensive agreement so difficult to achieve?

While not formally part of the UNFCCC process, countries can be grouped according to common economic and climate-related interests. Box 3.4 outlines some of the commonly recognised informal groupings. These groupings play influential roles in shaping climate negotiations. Crucially, these groupings can have very different perspectives on who is responsible for the climate problem, and who should therefore have the burden of both making the biggest reductions in GHG emissions, as well as providing finance and technology for poor countries to decarbonise and adapt to a changing climate.

It has long been recognised that different perspectives from different nation-state governments and associated groupings were impacting climate negotiations and thus the outcomes of the UNFCCC process (e.g. Ringius et al., 2002). The crucial points of difference generally arose out of different perceptions of what would be a fair outcome in relation to the burden of emissions reduction and financial and

Box 3.3 Climate finance under the UNFCCC

Key financial commitments

- In the Paris Agreement, developed countries pledged to *mobilise $100 billion annually* by 2020 to assist developing nations
- At *COP26 (2021), this goal was extended to 2025* as the target was not fully met
- At *COP28 (2023), negotiations began for a new post-2025 climate finance goal*, which will likely exceed $100 billion

Examples of country contributions

1. The **European Union (EU) and Member States**
   - *€28.6 billion in 2023* in public climate finance
   - Largest contributor to climate adaptation projects
2. The **United States**
   - Pledged $3 *billion* to the *Green Climate Fund (GCF)* (but has only delivered *$1 billion*)
   - Contributed *$9.5 billion in 2022* for international climate finance
   - Faces criticism for *falling short compared to its GDP size*
3. **Germany**
   - Contributed *€6.3 billion in 2022* (increased pledge at COP26)
   - Funds projects focused on renewable energy, water conservation, and resilience-building
4. **France**
   - Pledged *€6 billion annually* for climate finance
   - *One-third of its contributions* go to adaptation projects in vulnerable nations
5. **Japan**
   - *$1.5 billion pledge to the GCF*
   - Provides *climate-related loans* and technical support to Asia-Pacific nations

Recent developments and challenges

- **OECD Report (2023)**: Climate finance surpassed *$115.9 billion in 2022*, but *grants remain low* compared to loans

- The **Trump Administration's withdrawal** of support for the Paris Agreement in June 2017 and again in January 2025
- **UK Budget Cuts (2024–2025)**: The UK reduced its aid, impacting climate commitments
- **China's Role**: While not an official contributor, China provides *bilateral green investments* to developing nations

technology transfers between wealthier and poor nations. Salient amongst these different perceptions of fairness are:

- *The egalitarian approach*—where all people have an equal right to pollute or to be protected from pollution. In this case, emissions targets relate to emissions per capita where countries like the USA and oil-rich Middle East would have the biggest burden due to their high emissions per capita, while nation states with relatively high emissions but large populations (e.g. India and China) and the poorer, low-emitting states would have less of a requirement to act
- *The horizontal approach*—the view that the burden-sharing responsibility should be organised around countries with similar economic profiles, so that bigger economies have the biggest burden. In essence, this therefore considers the net cost of emissions reductions as a proportion of GDP. This would place the greatest burden on the largest economies including the USA, China, Japan, and Europe
- *The vertical approach*—the view that GDP size alone does not necessarily equate to the ability to pay because of population size in relation to GDP. For instance, while the USA and China may have similar sized economies, the wealth is distributed across a far great population in China than in the USA. From this perspective, therefore, the view is that the burden should be distributed according to the ability to pay, as determined by GDP per capita. This would mean that the burden would be principally borne by more mature economies in Europe, North America, Japan, and Oceania, as well as by oil-rich Middle Eastern states, whose GDP per capita are amongst the highest in the world
- *The polluter pays principle* stipulates that the burden of emissions reductions should be proportional to the level of emissions (possibly including historic emissions) of a country. This would mean that high-emitting countries such as the USA and China which, combined, emit well over a third of all global annual GHGs, would be required to make the biggest emissions reductions. If historical emissions are also accounted for, European states and Russia would also be required to take on a significant burden of GHG reductions

The Kyoto Protocol, with its focus on wealthier per capita Annex I countries bearing the largest burden of emissions reductions, was very much designed along the lines of vertical fairness principles, with some recognition of historical emissions, thus incorporating aspects of the polluter pays principle approach. While the Paris Agreement asks all UNFCCC signatories to devise NDCs for emissions reductions,

Box 3.4 The most prominent negotiating groups in UNFCCC negotiations

1. **G77 + China**
   - Represents *over 130 developing countries* (largest bloc)
   - Advocates for *climate finance, technology transfer, and equity* (Common but Differentiated Responsibilities—CBDR)
   - Diverse membership includes both *high-emission* (China and India) and *climate-vulnerable* (LDCs and AOSIS) countries

2. **The European Union (EU)**
   - Negotiates as a bloc of *27 member states*
   - Pushes for *ambitious climate action, emissions reductions, and global leadership on climate policy*
   - Played a key role in securing the *Paris Agreement*

3. **Umbrella Group**
   - A coalition of non-EU *developed countries: the U.S., Canada, Australia, Japan, New Zealand, Norway, etc.*
   - Generally *opposes strict climate commitments* and favours *market-based solutions*

4. **Least Developed Countries (LDCs)**
   - Represents the *poorest and most vulnerable nations*
   - Focuses on *adaptation finance, loss and damage, and capacity building*

5. **Alliance of Small Island States (AOSIS)**
   - Composed of *low-lying island nations* (e.g. Maldives, Fiji, and Tuvalu)
   - Advocates for *1.5°C warming limit, loss and damage compensation, and urgent action*

6. **High Ambition Coalition (HAC)**
   - A *cross-regional* group including *the EU, LDCs, AOSIS, and some Latin American countries*
   - Pushes for *strong climate targets and net-zero commitments*

this still embodies a vertical rationale, whereby those who have the most wealth per capita should do more, and also acknowledges the importance of wealth and technology transfers from the wealthier countries to the less wealthy—see Box 3.3.

Crucially, the preferences of UNFCCC nation-state participants are clearly visible in their negotiating positions. These have remained relatively fixed over time, albeit with some fluctuations occurring with geopolitical shifts. According to Hermwille et al. (2017), UNFCCC participants mirror global power imbalances. Indeed, the UNFCCC tends to be dominated by the differing views on what is a fair outcome between the world's two largest emitters and economies, China and the USA. These differences are fuelled in part by their economic competition as the two countries with the largest GDP share in the world. This rivalry has increased over recent decades as China has become wealthier, with a concurrent increase in CO_2 emissions. China's view is that, as a developing country with a large population, its responsibility for reducing emissions should be considered based on its GHG emissions per capita, which are relatively low by North American and European Standards—the vertical approach. Moreover, despite its high-volume emissions, China also argues that responsibility for reducing emissions should also be based on historical emissions (an aspect of the polluter pays principle), which again would suggest that America and Europe should primarily bear the burden of emissions reductions targets (Hurri, 2020). This is not to say that China has not drawn up domestic emissions reduction targets as part of the Paris Agreement, but that it argues that the biggest cuts should be borne by those with the broadest shoulders who have benefited most historically from burning fossil fuels (Hurri, 2020). By contrast, the USA position is that China, as the world's largest emitter of GHGs and second largest economy, should be on a level playing field with the USA in terms of bearing the burden of emissions reductions (Wu et al., 2022)—the egalitarian and horizontal approaches to fairness.

These opposing positions have been a real stumbling block in UNFCCC negotiations. The USA position has also been influenced by changes in the political make-up of the US Congress, which has led periodically to more hostile attitudes to climate action as discussed earlier. China's position has been broadly supported and shared by countries in a similar position, such as India and Brazil. Europe (the European Union and the United Kingdom) and Japan, in the meantime, accept historical and current responsibility for climate change and have been leading the way in terms of climate policy, alongside compensation and technology transfer to the most impacted countries.

For example, Japan's Joint Crediting Mechanism helps developing countries adopt low-carbon technologies while allowing Japan to earn carbon credits. Under this scheme, Japan has introduced energy-efficient inverter air conditioners that consume 30–40% less energy than traditional models in Indonesia, Vietnam, Thailand, and the Philippines. Moreover, Europe and Japan are amongst the biggest contributors towards the $100 billion finance goal from the Paris Agreement—see Box 3.2. Other countries' positions are shaped by their exposure to climate impacts, their vulnerability to climate impacts, their wealth and their relationships with key global actors, such as the USA and China (see Box 3.3), on power alignments and groupings in the UNFCCC. Key among these groups is the Organisation of Small Island States (with members from all over the globe including Antigua, Barbuda, the Bahamas, and Cook Islands) for whom climate change represents an existential risk, due to sea level rise (Canagarajah, 2024). For these countries, their survival depends on requiring all large GHG emitters to significantly reduce GHG, regardless of possibility and wealth.

Other key drivers of different negotiating positions have been the development of regulations linked to green technologies and associated capacity and capability. Front runners in these areas (in terms of deployment and in terms of production) have been pushing for solutions, for example, around renewable technologies. An illustrative example is the mass deployment of solar power in Germany through its innovative feed-in tariff (a guaranteed price for solar power exported to the grid) in the early 2000s and associated cost reduction of the technology due to mass market production in China, which has allowed for the wider deployment of this technology (Kuyper et al., 2018). However, different countries and regions have started to react negatively to technological developments, with associated impacts on climate goals and policy due to fears around economic competition. A salient example here is US and European reactions to the influx of cheap Chinese electric cars, which have put Chinese manufacturers at an advantage over traditional US and European car manufacturers. This situation has led to threats around the introduction of tariffs (on the back of alleged subsidies by the Chinese Government), which has taken on heightened significance in light of President Trump's strong focus on tariffs from the outset of his second term in office. Moreover, concerns over Chinese electric car manufacturing have led, particularly in Europe, to the moving of targets on the phasing out of combustion engine technology further into the future to allow domestic manufacturers a longer timeline to transition.

The evolving engagement of other, non-national government organisations in climate governance

While the key organisational stakeholders for the UNFCCC are the Governments of 197 signatory and thus voting nations, there are also a number of so-called official observer groups. Observer organisations do not have voting rights but play a crucial role in supporting the UNFCCC negotiations through providing expertise, policy advice, and facilitating dialogue between different stakeholder groups and hence fostering global cooperation in the fight against climate change (Hermwille et al., 2017). The UNFCCC classifies observer organisations into three broad categories:

1. **United Nations Agencies and Intergovernmental Organizations (IGOs)**—such as the World Bank, the United Nations Environment Programme (UNEP), and the World Meteorological Organization (WMO)
2. **Non-Governmental Organisations (NGOs)**—including environmental groups (e.g. Greenpeace and World Wide Fund for Nature (WWF)), business and industry groups, labour unions, and research institutions
3. **Constituency Groups**—these represent specific interest groups, such as youth, indigenous peoples, trade unions, and women and gender organisations

The increased recognition of the need to bring in different types of organisations to fully address the climate problem has grown over the years of the UNFCCC from just local governments in the first 1995 COP, to business NGOs, faith groups, farmers, and international organisations, such as the World Bank, among others in more recent COPs (Kuyper et al., 2018).

According to Kuyper and colleagues (Kuyper et al., 2018), the Paris Agreement achieved a particular landmark because of the importance it places on the role of non-state actors for meeting climate change targets, particularly in relation to national targets. This is a significant move away from the Kyoto Protocol which was far more state-centred. Indeed, since the Paris Agreement, climate governance has incorporated a greater diversity of actors operating at different scales of governance (see also later), with the understanding of the need for a more comprehensive and multilateral approach to achieve the goals of the Paris Agreement. The UNFCCC has also established several coordination platforms to include non-state and sub-national actors in climate governance (Kuyper et al., 2018). While these are not part of the formal decision-making process of the UNFCCC, they do bring perspectives from policy-facing organisations beyond nation-state governments to create opportunities for collaboration around finance and knowledge sharing—for more details on these platforms, see Box 3.5.

Box 3.5 The UNFCCC coordination platforms, which facilitate global climate action through collaboration, finance, and knowledge sharing

- **Marrakech Partnership (MPGCA)**—Engages governments and non-state actors to enhance climate ambition
- **NDC Partnership**—Supports countries in implementing Nationally Determined Contributions (NDCs)
- **Local Communities and Indigenous Peoples Platform (LCIPP)**—Integrates traditional knowledge into climate action
- **Technology Mechanism (TEC & CTCN)**—Promotes climate technology transfer
- **Adaptation Committee and Nairobi Work Programme (NWP)**—Focuses on climate adaptation strategies
- **LDC Expert Group (LEG)** and **National Adaptation Plans (NAPs)**—Supports Least Developed Countries (LDCs)
- **Climate Finance Platforms**—Includes the Standing Committee on Finance (SCF), Green Climate Fund (GCF), and Global Environment Facility (GEF)
- **Global Stocktake (GST) and Enhanced Transparency Framework (ETF)**—Tracks climate progress and transparency

Crucially, therefore, the evolution of international climate governance has created the space for transnational partnerships between organisations other than nation-state governments, including sub-state governments across the world, which are playing a vital role in addressing climate change through various strategies and policies (Wellstead & Stedman, 2011). These efforts are particularly significant given that many national governments have yet to implement strong climate policies or are grappling with differing political pressures. Sub-national entities, such as cities, municipalities, regional and state governments, and local districts, have the flexibility to implement tailored policies to promote sustainability, reduce emissions, and adapt to the impacts of climate change. Moreover, many regional and state governments have the authority to implement policies that can surpass national targets, offering a wider set of options to address climate change, setting ambitious renewable energy goals, and putting a price on carbon emissions to be used for setting policies (e.g. carbon taxes) to incentivise the adoption of lower carbon technologies.

Indeed, several regions have implemented carbon pricing mechanisms to reduce emissions. For instance, California, USA, operates a carbon trading scheme for heavily polluting industries whereby industries are allocated a carbon allowance which means any surplus can be traded with other industries that need to exceed their initial allowance. In this way, politicians can set a limit on the amount of GHG emissions from industry through the allocation of allowances, while also providing some flexibility for industries through allowing the buying and selling of excess credits resulting from industries adopting lower carbon practices and technologies. This scheme has been widened through being linked with a similar scheme in Quebec, Canada (Wolinsky-Nahmias, 2014). British Columbia in Canada introduced North America's first carbon tax in 2008, becoming a pioneer in pricing carbon to incentivise emissions reductions. Bavaria in Germany offers regional incentives to promote renewable energy production.

Moreover, urban areas, governed by local authorities, are often at the forefront of climate action due to their concentrated populations, industries, and energy consumption patterns. Cities have the ability to implement policies that affect transportation, energy use, waste management, and resilience to climate change. For example, Copenhagen in Denmark aims to become carbon neutral by 2025, setting an example for other cities worldwide. Vancouver in Canada aims to have 100% renewable energy by 2050, with a focus on clean energy systems and green infrastructure. Cities are also focusing on decarbonising transportation systems, which is a significant source of emissions. Oslo, Norway, plans to eliminate all fossil fuel emissions from its transport sector by 2030, while London, UK, has introduced the Ultra Low Emission Zone (ULEZ) to curb air pollution and encourage the adoption of electric vehicles (EVs). Paris, France, has banned diesel cars from entering the city and has created extensive cycling infrastructure to encourage sustainable modes of transport, and Bogotá, Colombia, is expanding its electric Bus Rapid Transit (BRT) system, a low-carbon alternative to traditional public transportation.

In addition to local and regional initiatives, numerous cities and regions are collaborating across borders to meet shared climate goals. These coalitions amplify the efforts of individual jurisdictions and accelerate the global transition to sustainability. Several networks of cities have been formed to collectively address climate change. Prominent among these, for instance, is the C-40 group, a global network of nearly 100 major cities committed to addressing climate change through bold action. Led by a rotating chair elected on a three-yearly term, it brings together city leaders to collaborate on policies and projects that reduce GHG emissions to achieve the goals of the Paris Agreement at the city level and to improve urban sustainability. Its member cities represent over 700 million people and one-quarter of the global economy, providing significant support for initiatives in areas like clean energy, air quality, waste management, and transportation. C-40 thus promotes city-level climate action by sharing best practices, providing technical assistance, and advocating for stronger climate policies worldwide. The ICLEI—Local Governments for Sustainability is another regional initiative that consists of more than 2,500 cities worldwide, working together to advance sustainable policies and climate action, while the Carbon Neutral Cities Alliance (CNCA) includes cities that are committed to achieving carbon neutrality and sharing strategies for success. Such regional pacts help align policies and pool resources to address climate change at a larger scale.

International climate policies are also shaped by companies and employer organisations (Bäckstrand et al., 2017). For example, over time, trade associations and trade unions have been given an increased focus and role in the UNFCCC process (Thomas, 2021). Whilst there is some debate about whether trade unions are conflicted by the 'jobs versus climate' discourse, that is, the perceived clash between preserving jobs and implementing climate change policies (Räthzel & Uzzell, 2011), Hampton (2018) notes a growing understanding on the part of UNFCCC and national governments that individual employers and workers have important voices and contributions to make around climate action and in determining the shape and nature of the green industrial transition (Stevis & Felli, 2015).

Alongside this trend, national (and trans-national) corporations are taking a more proactive stance on organisational decarbonisation. Many of the world's largest corporations have committed to future 'net-zero' emissions targets to meet shareholder, board, worker, and wider societal interest. This *political* relationship is key to driving decarbonisation (Downie, 2017), primarily by demonstrating the viability and benefits of a Net Zero business model, providing data and insights to support policy changes. At the *organisational* level, this happens through various eco-initiatives and programmes, but also through transforming their own internal structures, processes, and cultures towards climate action. Purpose-led organisational change, for example, is one core strategy by which corporations and firms are driving down emissions and contributing to more sustainable business practices. We discuss this further in the remaining chapters of the book. *Individuals* also play a crucial role, not only through employees seeking to promote climate action in an effort to be part of a sustainable future but also through leadership and social management. Indeed, prominent business leaders such as Bill Gates have become vocal advocates for fighting climate change, representing a range of different strategies, for example, business-led technological innovation (Gates, 2021), financial mechanisms, and green human resources management (Renwick et al., 2013).

Conclusion

As 'modern' society has become increasingly globalised and interconnected/interdependent, policy-facing organisations have evolved to respond to and help shape responses to contemporary global challenges such as climate change. This account helps us make sense of the, at times, bewildering organisational ecosystem of contemporary climate governance. We have drawn this account to try and bring some clarity to the practitioner community. Transnational climate governance could not exist and endure (let alone seek to address climate breakdown) without this complex ecosystem of multi-level, transnational, and diverse array of organisations representing varying interests and positions. Moreover, the trajectory of climate governance in the 21st century is arguably towards greater organisational complexity in form and substance where ratcheting climate action demands greater involvement and action from an ever-wider array of organisational interests and actors. As more and more global organisations are impacted by and hold positions relative to the climate crisis, so too will this organisational ecosystem become complex, crowded, and risk/opportunity laden. This situation generates associated challenges for achieving a meaningful international global climate agreement.

Further reading

Gates, B. (2021). *How to avoid a climate disaster: The solutions we have and the breakthroughs we need*. Vintage.

Harris, P. G. (2021). *Pathologies of climate governance: International relations, national politics and human nature*. Cambridge University Press.

Kirton, J. J., Kokotsis, E., & Warren, B. (2022). *Reconfiguring the global governance of climate change*. Routledge.

Sullivan, R., & Gouldson, A. (2020). *Climate change and the governance of corporations: Lessons from the retail sector*. Routledge.

https://www.worldbank.org/en/programs/climate-change-governance/blogs

References

Abbott, K. W. (2014). Strengthening the transnational regime complex for climate change. *Transnational Environmental Law*, *3*(1), 57–88. https://doi.org/10.1017/S2047102513000502

Bäckstrand, K., Kuyper, J. W., Linnér, B. O., & Lövbrand, E. (2017). Non-state actors in global climate governance: From Copenhagen to Paris and beyond. *Environmental Politics*, *26*(4), 561–579. https://doi.org/10.1080/09644016.2017.1327485

Bevacqua, E., Schleussner, C. F., & Zscheischler, J. (2025). A year above 1.5°C signals that Earth is most probably within the 20-year period that will reach the Paris Agreement limit. *Nature Climate Change*, 1–4. https://doi.org/10.1038/s41558-025-02246-9

Biermann, F., Siebenhüner, B., & Schreyögg, A. (Eds.). (2009). *International organizations in global environmental governance* (Vol. 17). Routledge.

Bulkeley, H. (2014). *Transnational climate change governance*. Cambridge University Press. https://doi.org/10.1017/cbo9781107706033

Canagarajah, S. (2024). Four things you should know: Climate change & Small Island developing states—by the numbers. *World Bank Blog*. https://blogs.worldbank.org/en/climatechange/four-things-you-should-know--climate-change---small-island-devel

Downie, C. (2017). Business actors, political resistance, and strategies for policymakers. *Energy Policy*, *108*, 583–592. https://doi.org/10.1016/j.enpol.2017.06.018

Gates, B. (2021). *How to avoid a climate disaster: The solutions we have and the breakthroughs we need*. Vintage.

Hale, T. (2020). Transnational actors and transnational governance in global environmental politics. *Annual Review of Political Science*, *23*(1), 203–220. https://doi.org/10.1146/annurev-polisci-050718-032644

Hampton, P. (2018). Trade unions and climate politics: Prisoners of neoliberalism or swords of climate justice? *Globalizations*, *15*(4), 470–486. https://doi.org/10.1080/14747731.2018.1454673

Hermwille, L., Obergassel, W., Ott, H. E., & Beuermann, C. (2017). UNFCCC before and after Paris–What's necessary for an effective climate regime? *Climate Policy*, *17*(2), 150–170. https://doi.org/10.1080/14693062.2015.1115231

Hoppe, R., Wesselink, A., & Cairns, R. (2013). Lost in the problem: The role of boundary organisations in the governance of climate change. *Wiley Interdisciplinary Reviews: Climate Change*, *4*(4), 283–300. https://doi.org/10.1002/wcc.225

Hurri, K. (2020). Rethinking climate leadership: Annex I countries' expectations for China's leadership role in the post-Paris UN climate negotiations. *Environmental Development*, *35*, 100544. https://doi.org/10.1016/j.envdev.2020.100544

Keohane, R. O., & Victor, D. G. (2011). The regime complex for climate change. *Perspectives on Politics*, *9*(1), 7–23. https://doi.org/10.1017/S1537592710004068

Kuyper, J., Schroeder, H., & Linnér, B. O. (2018). The evolution of the UNFCCC. *Annual Review of Environment and Resources*, *43*(1), 343–368. https://doi.org/10.1146/annurev-environ-102017-030119

Lieven, A. (2021). *Climate change: The greatest national security threat to the United States*. Quincy Institute for Responsible Statecraft. October 25. https://quincyinst.org/research/climate-change-the-greatest-national-security-threat-to-the-united-states/#executive-summary

Räthzel, N., & Uzzell, D. (2011). Trade unions and climate change: The jobs versus environment dilemma. *Global Environmental Change*, *21*(4), 1215–1223. https://doi.org/10.1016/j.gloenvcha.2011.07.010

Renwick, D. W., Redman, T., & Maguire, S. (2013). Green human resource management: A review and research agenda. *International Journal of Management Reviews*, *15*(1), 1–14. https://doi.org/10.1111/j.1468-2370.2011.00328.x

Ringius, L., Torvanger, A., & Underdal, A. (2002). Burden sharing and fairness principles in international climate policy. *International Environmental Agreements*, *2*, 1–22. https://doi.org/10.1023/A:1015041613785

Singh, S., & Jayaram, R. (2021). Climate finance: A systematic literature review. *International Journal of Economic Policy in Emerging Economies*, *14*(5–6), 566–580. https://doi.org/10.1504/IJEPEE.2021.118977

Stevis, D., & Felli, R. (2015). Global labour unions and just transition to a green economy. *International Environmental Agreements*: *Politics, Law and Economics*, *15*, 29–43. https://doi.org/10.1007/s10784-014-9266-1

Thomas, A. (2021). Framing the just transition: How international trade unions engage with UN climate negotiations. *Global Environmental Change*, *70*, 102347. https://doi.org/10.1016/j.gloenvcha.2021.102347

Vogler, J. (2016). The UNFCCC regime. In *Climate change in world politics. Energy, climate and the environment*. Palgrave Macmillan. https://doi.org/10.1057/9781137273413_3

Victor, D. G., & Keohane, R. O. (2010). *The regime complex for climate change*. APSA 2010 Annual Meeting Paper.

Wellstead, A. M., & Stedman, R. C. (2011). Climate change policy capacity at the sub-national government level. *Journal of Comparative Policy Analysis: Research and Practice*, *13*(5), 461–478. https://doi.org/10.1080/13876988.2011.605937

Wolinsky-Nahmias, Y. (Ed.). (2014). *Changing climate politics: US policies and civic action*. CQ Press.

Wu, Z., Huang, X., Chen, R., Mao, X., & Qi, X. (2022). The United States and China on the paths and policies to carbon neutrality. *Journal of Environmental Management*, *320*, 115785. https://doi.org/10.1016/j.jenvman.2022.115785

4 The shift from shareholder capitalism to marginalised stakeholder capitalism

Rashedur Chowdhury

Key points

- While traditional stakeholder management approaches represent an advance on shareholder capitalism that affords primacy to profit maximisation, this does not go far enough as a foundation for purpose
- The notion of marginalised stakeholder capitalism is put forward as the basis for challenging and transforming traditional shareholder and stakeholder capitalist frameworks
- Ten capabilities can be identified for marginalised stakeholders that enhance inclusivity and equity and that can be used to inform purpose development
- Individual, organisational, and societal well-being can be improved through marginalised stakeholder capabilities development
- Organisations have a key role to play in developing a purpose that addresses moral and ethical dilemmas in environmental and human engagement

Introduction

Shareholder capitalism is based on the premise that firms should focus solely on maximising profits to serve the interests of their shareholders. When firms are viewed through this lens, corporate managers are incentivised to disregard the interests of other stakeholders (e.g. customers, employees, suppliers, distributors, regulatory bodies, and nature) while maximising profits for shareholders. As a result, managers can take decisions that may not consider whether firm activities negatively affect the environment, employees, or broader society. At the same time, managers also have an interest in limiting the scope of regulations and restrictions so that they can avoid any legal or societal consequences for their exploitative or profit-maximising activities.

Ideally, shareholder capitalism thus prefers a laissez-faire approach which would help firms to take control over the market, set prices, and control the quality of products and services through competition. These perspectives have been fortified since Friedman (1970), a notable American economist, argued that the only responsibility of business is to maximise profit for shareholders. By doing so, businesses forgo social and environmental concerns as the argument is that these should be left

DOI: 10.4324/9781003545224-5

to governments and individuals. Friedman's ideas were influential as he was a leading advocate of neoliberalism, promoting free markets and minimal government intervention. His views aligned with the deregulation policies advanced by President Reagan in the US and Prime Minister Thatcher in the UK during the 1980s, as we saw in Chapter 3. These encouraged both institutional and private investors to focus solely on short-term financial performance, which then became a dominant market standard. This was reflected in how executive compensation was tied to stock performance, emphasised by cost-cutting measures to increase efficiency at the expense of environmental, social, and ethical performance.

In contrast to shareholder capitalism, stakeholder capitalism is based on the premise that firms should take care of a broad range of stakeholders, including employees, customers, suppliers, communities, and the environment. Though the notion of stakeholder capitalism has existed for decades, R. Edward Freeman's (1984) pioneering work on stakeholders, also known as stakeholder theory, challenges the core principles of shareholder capitalism. In the years that followed the publication of Freeman's work, highly controversial corporate scandals, such as American energy firm Enron's bankruptcy due to major accounting fraud and the 2007 financial crisis, created momentum for the increasing turn to stakeholder capitalism.

Under the broad umbrella of stakeholder capitalism, corporate social responsibility (CSR) and stakeholder management are two important perspectives that set out to challenge the shareholder capitalism on which many Western firms still base their socio-economic activities, feeding into organisational purpose. Friedman's ideas have been vigorously contested within the CSR and stakeholder management literatures (Freeman, 1984; Gabaldon, 2006; Sheehy, 2005), which instead argue that a firm's responsibility towards its stakeholders and the wider environment is just as important as shareholder value creation.

Although the traditional stakeholder management literature has questioned the dominant shareholder value-creation perspective (Dodd, 1932; Freeman, 1984; Gabaldon, 2006; Sheehy, 2005), in this chapter, I argue that the typical debates—such as shareholders versus stakeholders, or economic versus social perspectives—still do not go far enough in explaining the environmental and human degradations that occur under the traditional capitalist model. This is because both shareholder and stakeholder capitalism do not position marginalised stakeholders (MSs), including the environment, as the primary stakeholders of firms, or as the primary focus of core strategy, purpose, and firm-level routines. This is particularly important in the context of the current climate crisis and the issue of climate justice, wherein MSs are often most vulnerable to the effects of global warming despite being least responsible for it (the differential impacts of climate change are discussed in more detail in Chapter 2).

I define MSs as groups and individuals who are generally perceived as powerless or vulnerable. More specifically, MSs can be referred to as those who come from lower social or working classes, or may possess vulnerable social identities based on gender, race, sexuality, ability, age, immigrant status, political affiliation, or those in positions that limit their ability to advocate for themselves or lead change (Chowdhury, 2023).

Crucially, in the context of this book, the natural environment itself can also be considered as an MS, because organisations repeatedly use natural resources from the environment, whether directly or indirectly, without nature being able to protest or voice concern when it is being negatively affected. Instead, it relies on other MSs, such as indigenous populations and grassroots activists, to speak out on its behalf with the level of urgency required to protect the interconnected interests of planet and humanity. If firms are unaware of or ignore their MSs, or do not take environmental harms seriously, then corporate decisions end up being taken without any input from those MSs. In this context, any governmental policy to protect the environment or MSs may be ineffective. For example, firms can often find loopholes in government policies that are supposed to protect MSs. But it can be difficult for firms to know precisely how MSs can be affected by these loopholes if they are unwilling to consult the MSs in the first place. Therefore, strong government policies or regulations can work effectively if firms are willing to coordinate between policies and different stakeholder groups, including MSs.

From this perspective, both in the traditional business literature (including among proponents of stakeholder capitalism) and practitioners, there is often the mistaken assumption that firms can tackle social and environmental issues on their own, without government regulations or any meaningful in-depth engagement with MSs. In practice, many firms have a limited understanding—or even a lack of motivation—when it comes to addressing social and environmental issues without clear regulatory guidance or engagement with MSs. Strong laws, backed by representative bodies like parliament, and input from those MSs who are affected by environmental harm, are essential to ensuring businesses meet societal and environmental expectations. We discuss the global policy context for organisational action in more depth in Chapter 3.

Instead, we often see that firms use approaches such as organisational purpose, CSR, or stakeholder management under the guise of stakeholder capitalism as 'strategic tools' with which to navigate debates regarding environmental destructions or human miseries that are caused by the activities of the business and their associates. Another approach is the maintenance of alliances with different powerful actors, such as non-governmental organisations (NGOs), politicians, and government agencies (Chowdhury, 2021a). By 'strategic tools,' I mean that firms risk using purpose and stakeholder management approaches as greenwashing mechanisms. For example, some firms claim they source most of their products from fair trade but, in reality, source perhaps only 5–10% of their products from fair trade. In this case, they use the fair trade label as a strategic tool. Alternatively, they may make specific public claims (e.g. Laufer, 2003) with the help of powerful stakeholders or third parties, such as NGOs or government agencies, without directly involving themselves with the core recipients, that is, the MSs, who are directly affected by the firm's actions.

Therefore, I argue in this chapter that even stakeholder capitalism can often result in a short-term, bandaged solution to firm-specific problems. In the long run, it does not fix the core problems that ultimately have the greatest impact on MSs, wider society, and the environment.

Box 4.1 Asia Energy

Asia Energy, a mining firm, tried to acquire lands from the low-income population of Phulbari, Bangladesh (who we can consider as the MSs of Asia Energy) to operate an open pit mine. To acquire the land, Asia Energy initially used several NGOs to convince the Phulbari people that they should sell their lands to Asia Energy (Chowdhury, 2021a). When Asia Energy failed to convince the Phulbari to relinquish their lands, it used government machineries, such as law enforcement agencies, to inflict violence against them (Chowdhury, 2021a). In the process, Asia Energy made use of powerful (external) stakeholders, including NGOs and government agencies, to maximise shareholder value (i.e. to protect the interests of internal powerful stakeholders such as shareholders) at the expense of the Phulbari people. Asia Energy's engagement with powerful stakeholders and its actions in protecting the interests of shareholders not only violated the natural and representational rights of MSs, the entire process actually skewed the stakeholder approaches of Asia Energy towards a narrow economic value creation perspective (Chowdhury, 2021a; Falguni, 2009).

The Asia Energy example indicates that not only are firms' narrow economic approaches problematic but they also limit the ability to comprehend and address societal and environmental challenges. Even traditional stakeholder management approaches cannot effectively overcome the constraints of such narrow economic thinking since stakeholder engagement tends to be instrumental in nature. To do the right things morally and to pursue an authentic purpose-driven stakeholder engagement, firms need to cooperate with MSs in ways that are morally and ethically robust. Accordingly, a critical question arises: *How can firms fulfil their moral and ethical obligations towards MSs whilst supporting inclusive growth?*

Based on my research (Chowdhury, 2021a, 2023), I argue that, to assist firms in dealing with moral and ethical dilemmas in relation to societal and environmental challenges, we require radical and alternative perspectives that can move beyond the prevailing dominant theories and problematic (corporate or organisational) practices. There may be multiple ways to achieve this, but, oftentimes, we ignore the most obvious ways to 'see' and 'do.' For example, all types of firms are affected by, or affect, diverse sets of stakeholders and communities around them. It is generally assumed that, since firms cannot pay equal attention to all their stakeholders (Pfeffer & Salancik, 1978), they identify the most powerful ones and trust that they will take account of any relevant MSs. By doing so, firms (un)consciously ignore MSs. Such oversight is not only self-depriving—or even, in some cases, self-destructive—but also detrimental to firm performance and responsibilities because, if firms cannot recognise, comprehend, and actualise MSs' inputs (Chowdhury et al., 2024), firms may face protests, activism, or social sanctions from MSs. This (in)directly affects all stakeholders and parties because they are bound by complex, multilayered interrelationships.

Box 4.2 Shell in Nigeria

In the 1990s, Shell faced severe backlash in Nigeria due to its alleged involvement in the execution of nine Ogoni activists, including Nobel Peace Prize nominee Ken Saro-Wiwa (Boele et al., 2001a, 2001b). Years later, Shell paid $15.5 million in compensation to Saro-Wiwa's family for his death (Pilkington, 2009) and was also ordered by a U.S. court to pay $55 million for environmental damage in Nigeria (Vidal, 2015). The court found that oil spills and oil contamination had caused damage to ecosystems, including natural habitats for birds, fish, and wildlife, as well as harm to farmland, waterways, and drinking water sources. Apart from the financial losses involved, Shell's actions led to a severe crisis of legitimacy. Although the firm managed to survive, its reputation was deeply damaged in the eyes of society and MSs.

To this end, I offer the notion of *marginalised stakeholder capitalism* (MSC) as a new way for organisations to take MSs into consideration when developing their corporate purpose. I hope this can challenge the dominant and problematic elements of both shareholder and stakeholder capitalism and the bandaged solutions that firms often use to address the core problems underpinning purpose development. Through the concept of MSC, I emphasise a deeper, more direct, and cooperative engagement that can help us to revisit existing moral and ethical conceptions that are embedded into firm-level and societal functionalities.

Through the idea of MSC, I argue that the neglect of MSs in the age of social media and hyper-competition, geopolitical complexity, and fast-moving environmental catastrophic scenarios is impractical. This oversight is a concerning issue, as this can perpetuate human miseries and environmental destruction. To avoid this, my research shows firms must clarify or (re)establish their moral and ethical principles within their purpose statements so that they can address the problems that are affected by, or can affect, not only powerful stakeholders but also marginalised individuals and groups (Chowdhury, 2021b, 2023). There is an imperative for them to start seeing their activities, social responsibilities, and moral duties through the lens of MSs. By doing so, firms can develop strong moral and ethical claims and engage seriously with the concerns of societies and the environment. This can also help firms to find collective grounds on which to be truly conscious of the concerns of MSs and the various dilemmas they face.

For example, firms often acquire natural resources from MSs, such as indigenous groups, through violence or manipulation. However, neither violence nor manipulation is a legitimate way to engage or communicate with MSs. Sourcing natural resources requires the consent of MSs and meaningful discussions with them, which firms may fail to achieve. Consequently, a firm's engagement with MSs ceases as soon as it denies their legitimate representation. This legitimacy also extends to nature itself, which, being unable to communicate, remains inaccessible to firms. From this perspective, it is crucial to acknowledge the roles of local, regional, and national governments, as well as national and international agencies, alongside MSs in protecting it. For authorities and agencies that either enact laws and regulations or safeguard nature through non-regulative means, it is also important to consider MSs as a primary source of interaction and knowledge exchange. With the support of MSs, different governments and agencies can implement stronger legal and (non)regulatory frameworks that more effectively identify oversights and harmful actions by firms that damage nature.

Hence, communicating with and listening to the concerns of MSs is not an easy task—especially when (non)regulatory weaknesses or lack of enforcement persist due to the historical exclusion of MSs and, more specifically, when firms have long ignored them or denied their representational claims, including those made on behalf of the natural world. Nonetheless, I argue that firms ought to (re)organise or (re)develop the necessary organisational resources and routines that would enable them to break their path-dependent behaviour. In order to achieve this, I would offer the perspective of MSC which can be used as a lens and as a primary foundation to underpin and understand MSs. More importantly, I suggest that firms develop MS capabilities as a core intervention which can also serve as the basis for redeveloping organisational purpose and routines, and restructure firms' resources to achieve a stronger and more purposeful sociopolitical, environmental, and economic orientation.

The chapter is organised as follows. First, I analyse how MSs have been ignored in traditional debates around purpose and stakeholder management. Second, drawing on the expertise of scholars Sen (1999, 1985) and Nussbaum (2000), I argue that MSs can be cooperative partners for firms to acquire renewed—or new—organisational resources. Firms are then able to engage directly with MSs and conduct their business in more ethical and sustainable ways. Sen's 'capabilities' approach focuses on 'what people are effectively able to do and to be' (Robeyns, 2005, p. 94). Using this framework, I suggest how firms can develop MS capabilities that can enable MSs to represent themselves and their concerns in organisational and public debates and, thus assist firms to do things differently by ensuring equity and diversity. This means giving MSs greater voice and participation in firm decision-making, as their local knowledge and community awareness can prevent firms from sidestepping regulations designed to reduce human and environmental exploitation.

With more MSs involved in organisational and societal activities, firms are more likely to comply with socio-economic, environmental, and legal requirements, building the firm's reputation and ensuring inclusive growth. To effectively address today's environmental and human crises, and to develop an authentic organisational purpose, we must reimagine our capitalist system with MSs at its centre.

MSs in the shareholder–stakeholder debate

The central debate in the stakeholder–shareholder relationship is whether firms pursue certain objectives to satisfy shareholders for economic reasons or prioritise stakeholders alongside shareholders (Donaldson & Preston, 1995). Proponents of shareholder capitalism like to argue that firms hold multiple legal contracts between suppliers, distributors, and employees who are all involved in the production and sale of goods, and who agree to cooperate with one another to generate profitability for their owners or shareholders (Easterbrook & Fischel, 1991). These legal agreements specify that, in exchange for their services, the owners of production will be paid maximum profits. As a result, managers have a fiduciary duty to maximise shareholder value rather than switch their capabilities to social or environmental activities (Butler & McChesney, 1999). Viewed from this perspective, diverting efforts away from profit maximisation towards bolstering a socially or environmentally grounded organisational purpose could almost be viewed as theft of shareholders' money (Friedman, 1970). This is the argument that essentially underpins the view that managers can ignore stakeholder interests while focusing on maximising profits for shareholders.

However, proponents of stakeholder capitalism argue that the law permits managers to make decisions, if necessary, at the expense of shareholders in order to conform to social and moral obligations (Dodd, 1932; Sheehy, 2005). This is because corporate managers have fiduciary responsibilities to a wide variety of stakeholders, and, for social interest, it is acceptable to forgo shareholder profits. According to Freeman (1984), systematic managerial attention to stakeholder interests is critical to a firm's success. Hence, Freeman refers to a firm's stakeholders as not only its shareholders but also its customers, employees, suppliers, pressure groups, government entities, and local communities.

However, the idea of stakeholder capitalism does not provide managers with clear guidance on how to judge the relative importance of different stakeholder groups. Depending on the context, some stakeholder groups may need more attention than others. To aid decision-making, Mitchell et al. (1997) developed a theoretical model that managers can use to evaluate the salience of stakeholders based on the assumption that a stakeholder group can only influence managerial decisions if they accumulate three attributes: power, legitimacy, and urgency. However, this largely depends on managers' perceptions and their discretion in recognising stakeholders and addressing their concerns:

- **Power** arises when stakeholders have the ability to influence firms' decisions through sociopolitical, cultural, and economic means

- Whether stakeholders perceive a firm as a socially acceptable entity determines its **legitimacy**
- **Urgency** arises when an issue requires immediate attention to fulfil stakeholders' demands

If managers perceive that a stakeholder possesses only one or two attributes, they may not consider the stakeholder important or may even ignore them. This is especially true for MSs, as managers often do not view them as powerful or legitimate. Even when MSs face urgent issues, they are easily overlooked due to their lack of influence or authority.

One example of this is the Brent Spar case. Brent Spar was an oil storage and tanker loading buoy in the North Sea area of the Brent oilfield, which was owned and operated by Shell. When attempting to dispose of this facility at the end of its useful lifespan by sinking it in the North Sea, Shell initially failed to consider important environmental issues. If the storage facility and tanker were sunk, this could have damaged the deep-sea ecosystem and harmed marine wildlife. Shell later changed its decision due to Greenpeace activism. In this case, Greenpeace, as a well-known NGO, had leverage against Shell, but this is a considerable weakness of the saliency model in the sense that it relies on powerful stakeholders to prompt action (Zyglidopoulos, 2002). Indigenous populations or unknown activists often have less legitimacy or fewer resources at their disposal to highlight their concerns in protecting their lands or environment.

Mitchell's stakeholder saliency model is based on an 'instrumental stakeholder perspective' that does not consider (post)colonial contexts that affect the relationships between firms and MSs, and instead emphasises how MSs can help generate higher economic value (Chowdhury, 2021b, 2023). Examples of this include forced land grabs in indigenous territories or poor communities in Africa in relation to mining operations or illegal waste dumping, or neocolonial instrumentalities. Even in the USA, excessive and unregulated corporate involvement has made healthcare and housing unaffordable for ordinary citizens, leading to early deaths, homelessness, and increased crime (Angell, 2005 Desmond, 2016; Hacker, 2006.

Box 4.3 Rana Plaza

In this context, we can consider what happened when the Rana Plaza building collapsed in Bangladesh to help us understand why instrumental approaches are largely ineffective in solving contemporary societal problems. Thirty-one Western multinational clothing firms had been sourcing clothing from five garment factories situated in the eight-storey Rana Plaza building in Bangladesh. The building collapsed in 2013 due to structural failure, killing and injuring at least 1,135 and 2,500 workers, respectively. While the Western firms completely denied their responsibility for the collapse, in the aftermath of the disaster, two alternative institutions, Accord and Alliance, were built with backing from Western multinational clothing firms to ensure future building safety. However, both institutions mostly served the interests of Western firms at the expense of improving the well-being and representational rights of the local workers (Chowdhury, 2017a, 2023; Muhammad, 2015). In the end, those institutions were used as a strategic tool to ensure the appropriate discussions did not take place, and circumvent the proper distribution of compensation to Rana Plaza victims (Chowdhury, 2017a).

Given the type of disaster that happened and the number of people who were severely injured or died (Chowdhury, 2017b), one could estimate that if a legal action was lodged against these firms by the Bangladesh government, the Western firms might have ended up paying billions of dollars in compensation (Chowdhury, 2017a). But, due to weak institutional settings, including non-mobilisation of regulatory frameworks and strategic use of CSR—such as Western firms hiring NGOs to evaluate injuries and provide minimum compensation—a great injustice was caused to the workers who laboured day and night for these firms. This case illustrates how firms often adopt saliency-based approaches to engage NGOs or local elites—or exploit loopholes in legal and regulatory systems—to create institutions that deflect their direct socio-economic and legal responsibilities (Chowdhury, 2023). This often hinders marginalised workers from mobilising genuine representation or forming independent unions (Shifa et al., 2015)—beyond local political affiliations that do not truly reflect their interests (Taher, 1999)—especially in the absence of meaningful intervention by external institutions like Accord and Alliance, which themselves allowed hardly any representation of marginalised workers in this context (Chowdhury, 2023).

Moreover, in contemporary contexts, issues such as race, diversity, protests, and activisms are often ignored or excluded due to a reliance on the saliency approach (Chowdhury, 2021c), simply because MSs do not have enough power to claim legitimacy or urgency (Chowdhury, 2017a). As a result, even when issues such as social movements or protests are taken into consideration under the umbrella of the stakeholder perspective, the emphasis still remains on stock performance or profit maximisation (King & Soule, 2007), with MSs viewed as peripheral or insignificant.

From this perspective, when firms consider MSs, they not only see them in instrumental terms (e.g. Henisz et al., 2014) but also (un)consciously design and perform organisational activities that often violate the expectation of stakeholders and, thus, MSs' needs and expectations end up being neglected both in the stakeholder and CSR/purpose scholarship and in practice.

For example, in response to the Phulbari villagers' protest against Asia Energy's attempt to grab land, a former executive of the firm mentioned that

> they had no idea ". . . why these people were protesting. They were too poor, too uneducated. We wanted to mine Phulbari to give them better lives. They were denying themselves good lives. That made us laugh about these people."
>
> (Chowdhury, 2021a, p. 142)

Such intentional portrayal of MSs as undignified is not only insensitive (Chowdhury, 2021a) but also impractical for developing any constructive engagement aimed at achieving conflict resolution.

Even when firms do appear to consider the moral argument about how they should deal with their stakeholders, there is still the risk that they end up adopting a more instrumental approach based on their perception of the most efficient allocation of time and resources, which can favour powerful stakeholders at the expense of MSs. More problematically, this then results in firms further neglecting the need to recognise and develop the MS capabilities which could help them to (re)focus on the fundamental issues that impede them from dealing with human and environmental degradations and thus become authentically purpose-led.

Mobilising marginalised stakeholder capitalism

MSC (Chowdhury, 2013, 2021a, 2021b, 2023) can be mobilised in different ways within the organisation. Considering marginalised stakeholder *capabilities* for MSC is one of the ways firms can benefit. Sen (1999, 2005) first explained how capabilities are beneficial in helping individuals achieve freedom and skills so that people can choose what they want to do with their lives. In other words, Sen's idea of capabilities refers to social arrangements that should provide ordinary citizens with the necessary opportunities to leverage their empowerment to make choices that are meaningful to them.

From this perspective, in development economics (which is a branch of economics that focuses on economic transformation while also addressing issues such as poverty, human development, and structural change), capabilities are not just tools to attain goals such as sustainability or fairness. Instead, they are important goals in their own right. This means that policies and projects—whether run by governments, businesses, or partnerships—should focus on building ordinary people's skills and opportunities as a key part of socio-economic development.

Therefore, when I use the term 'MS capabilities,' I use the term 'capabilities' in the sense that they would provide MSs with equity, self-representation, and freedom to live a decent and dignified life. For example, a woman should have the freedom to work even though a patriarchal society may create obstructions to her enjoying the life she wants to pursue. From this perspective, fundamentalism, extremism, or any other type of misogynistic or racial approaches should not hinder her socio-economic opportunities. The idea of individual freedom (a type of freedom that does not hinder or intimidate individuals, directly or indirectly, from revealing their identity, expressing their opinions, and pursuing their legitimate right to live with dignity in both work and non-work contexts) is applicable at the group or community level and can be aggregated to societal freedom. If individuals have freedom, then groups and societies will naturally be free as well. But if individuals lack freedom, it means society is not truly fair or inclusive, especially for minorities. Interestingly, once we conceptualise MS capabilities as worthwhile ends in themselves towards which firms can strive in their organisational purpose, we can see firms start creating better ways of engaging with their day-to-day activities.

For instance, when a firm operates in a deprived or postcolonial environment and strives to embrace and enhance the development of MS capabilities, such as improved health and job skills for people in those communities, it stands to gain not only a better labour force for the future but also a labour force emotionally vested in a mutual path for socio-economic success. Thus, the idea of developing MS capabilities is to enrich firms' routine structures and organisational resources. These can then actualise those MS capabilities that firms choose as ends, not simply as rhetorical means to fulfil the stakeholder commitments embedded in their published purpose. For example, pharmaceutical firms often claim to invest in their purpose by funding NGOs to carry out social initiatives. However, this approach allows firms to remain hands-off rather than directly engaging in efforts to benefit society. If these firms instead developed innovative practices to actively support marginalised communities, they could contribute more effectively to societal progress.

MS capabilities as ends

I propose ten MS capabilities drawing on Nussbaum (2000). These can provide the basis for guiding managers in actualising their commitments to MSs, and can be further developed and expanded to identify other MS capabilities and sets of capabilities (which can be country, industry, or firm-specific). These can then underpin the organisational purpose. Each of these MS capabilities (marked in italics) is shown in Table 4.1 and is briefly described next.

- **Physiological:** At the very least, MSs want to be paid a dignified living wage so that their physiological needs are fulfilled. This means that MSs do not need to beg anyone for a minimum living standard that they deserve as a human being. It can give them confidence to work for an organisation which at least takes care of the basic needs of MSs. This refers to *physiological MS capabilities.* Physiological needs are the physical requirements for human functioning (Maslow, 1943), for instance, food, clothing, and shelter are basic elements that are required for decent survival. Often, MSs are deprived of these basic needs even when they are employed due to low wages and exploitative job conditions. For example, trafficking of humans can be considered to be extremely

Table 4.1 Marginalised Stakeholder capabilities.

Core capabilities	*Translated into MS capabilities*	*Firms' perspective on MS capabilities development*
1. Life	*Fulfilling physiological needs*	Fulfil the basic needs of MSs
2. Bodily health	*Healthiness*	Make MSs physically and psychologically sound
3. Bodily integrity	*Integrity*	Ensure mobility of MSs
4. Senses, imagination, and thought	*Equity*	Enable MSs to make fair judgement and decisions
5. Emotions	*Affective*	Allow MSs different types of emotional expression
6. Practical reason	*Rationality*	Treat MSs as rational partners
7. Affiliation	*Recognition*	Protect the dignity of MSs
8. Other species	*Sharing*	Share an open-minded attitude with MSs to live with others in harmony
9. Play	*Entertainment*	Allow MSs to derive happiness from simple things
10. Control over one's environment	*Political and material control over one's life*	Allow MSs freedom of expression, voting rights, and trade union associations

exploitative as MSs may rely on their captors for their basic needs. Mining areas that are prone to conflict can also affect MSs adversely. Such conflict not only violates MSs' rights but also the rights of nature not to be disturbed or harmed, as MSs are often the primary custodians of nature. It can even be the case that a community is marginalised due to an oppressive state where domestic or foreign firms take the opportunity to utilise cheap labour and design employment conditions that automatically exploit the workers or their surrounding communities. Thus, firms must actively challenge and overcome oppressive conditions to ensure that the development of MSs' physiological capabilities is not hindered.

- **Healthiness:** When MSs meet their physiological needs, they demand good health. *Healthiness MS capabilities* are important because good health ensures that a stakeholder functions effectively in the workplace without long-term negative health implications. Moreover, a firm's provisions for instances of ill health among their workers or members of their supply or distribution networks can provide a stakeholder with security.
- **Integrity:** *Integrity MS capabilities* mean that MSs can move from one place (e.g. different geographic areas, communities, and workplaces) to another. If MSs are locked into a specific space and cannot move into other communities or organisations due to a deficiency in capabilities, this means that they have not been able to develop the necessary skills that are supposed to give them confidence and freedom to move freely and make life choices. For example, a social media platform has a responsibility to restrict content that may increase child and sexual abuse. If this means the social media platform needs to provide their marginalised customers with more protections, these platforms should

ensure that they do so. These platforms cannot expect vulnerable individuals to be responsible for firm behaviours.

- **Equity:** Providing stakeholders with *equity MS capabilities* is important. Equitable MS capabilities mean that MSs can make a sound decision about their lives, careers, and the environment, including the natural resources they use or protect for future generations. This also means that MSs are seen as equal partners of powerful stakeholders in and around the organisation and are integrated into the organisational or community culture. Hence, for instance, the organisational culture must be more inclusive so that all stakeholders can freely express their opinions and practice their religious beliefs, political views, and so forth. This may even mean that MSs form trade unions or other types of collective forums to express their concerns and demands (cf. DiNardo & Lee, 2004; Stevens, 1990).
- **Affective:** The development of *affective MS capabilities* is also important. Affect refers to various forms of emotional expressions or displays (Batson et al., 1992). These displays can take place through facial expressions, vocal expressions, or gestures. The accommodation of various affective displays helps MSs form human contacts, establish personal relationships, and express their feelings, thereby reducing anxiety and stress. This is also a mechanism to protect nature and the natural resources that MSs cherish and safeguard in the best interests of both humanity and nature itself. For example, if MSs protest against a mining firm to protect their lands and the surrounding environment, they should have the ability to show their legitimate emotions, such as concern or anger, in carrying out their activism. Such activism should be taken seriously by firms to accommodate affective MS capabilities.
- **Rationality:** *Rationality MS capabilities* mean that MSs are capable of undertaking practical reasoning. For this, MSs may use past experience or conscience to make logical decisions about their work and the exercise of their rights to protect themselves and the environment. Even if an MS has not received a formal education, this does not mean that they are politically unaware or emotionally unable to express political opinions (Chowdhury, 2021b; Nussbaum, 2013). For example, even without formal education, MSs, such as indigenous populations, have a deep understanding of nature and natural resources, enabling them to navigate their lives without harming the environment. Indigenous populations are also aware of the political implications, including the risk of violence, if they do not actively protect nature.
- **Recognition:** *Recognition MS capabilities* enable one to live and work with dignity. An MS should be protected from any humiliation that may occur because of their appearance, ethnicity, sexual orientation, religion, race, immigration status, nationality, and their reliance on the environment, which may provide them with peace and prosperity.
- **Sharing:** *Sharing MS capabilities* enable MSs to live in harmony with other species and the world of nature. Specifically, if a marginalised community is affected due to mining or other activities which could destroy the human and environmental harmony and the natural ecosystem, firms have the responsibility to either stop their operations or bring about significant levels of interventions to minimise disruptions and damages. For example, Asia Energy should stop

encroaching on the Phulbari region and its people to acquire land for mining. Since the people of Phulbari have rejected the idea of mining in their area, Asia Energy has a duty to respect and acknowledge their aspirations and expectations, rather than continually using manipulative techniques to obtain the land (see Box 4.1).

- **Entertainment:** *Entertainment MS capabilities* enable MSs to enjoy life while working. Simple things, such as the ability to laugh and gossip, enable MSs to enjoy their life within and outside firms. Enjoying nature in quietness or living a peaceful and simple life can also be a source of entertainment for some MSs (e.g. indigenous populations), and this must be respected.
- **Political and material control:** Finally, *political and material control MS capabilities* to have a rightful representation and a say over the MS environment are crucial. This enables stakeholders to make use of voting choices and property rights to protect personal belongings and common resources (e.g. forests, forest habitats, plants, woodlands, rivers, and ponds). MSs can understand these rights when they work for a firm that adheres to such rights. MSs are also a practical source of firm engagement with wider society and politics. In this sense, the input of MSs regarding their political and material views is important for the continuous equitable improvement of firms' ability to understand their external environment. By doing so, firms would become more sensitive to environmental issues and consider MSs from an environmental perspective. This is crucial because MSs and the environment often go hand in hand, and one cannot be substituted for the other.

Discussion

Through the notion of marginalised stakeholder capitalism, I argue that, by actively engaging with MSs, firms can better understand their concerns and expectations. When firms engage with MSs directly through mechanisms such as MS capabilities, firms can then (re)configure their routine structures to gather or update raw data and real-time information about MSs and can use this information to formulate a more effective MS engagement approach than is possible via the traditional stakeholder management approach.

Further, if firms are able to develop MS capabilities through a transparent process, they do not need to seek legitimacy and mandate from powerful stakeholders or third parties, such as government agencies and NGOs, in order to address the plight of MSs because the MSs can endow firms with intergenerational, historical, and relational opportunities to collectively achieve their rightful objectives. In other words, when MSs enjoy their rights and firms can accept any type of rejection or feedback from MSs who have intergeneration knowledge about the firms, then both have the opportunity to engage constructively by forgoing past disputes. Thus, from the perspective of marginalised stakeholder capitalism, firms cannot expect to win all the time.

Also, often it might be impossible for firms to create a win-win situation. A win-lose situation must be respected by firms where firms might be losers from an economic perspective but clearly can help MSs and the environment to win in order to achieve greater societal and intergenerational well-being. It is worthwhile pointing out that when firms lose investment or profits in certain parts of their business, they still have the ability to perform well in other areas. This is especially true for multinationals or large firms, which are often more responsible for environmental or other forms of damage. A true balance between winning and losing is essential for fostering inclusive growth—something that most firms operating under shareholder or stakeholder capitalism often overlook.

Engaging directly with MSs would equally create an opportunity to discuss and debate concerns and grievances that firms often do not accommodate through stakeholder capitalism. If firms can see the value of interacting and developing sets of capabilities with MSs, it means that firm-level policy formulation and implementation—attaining legitimacy and mandate from the MSs—can be a pathway to ensure the representational rights and dignity of the MSs, as well as broader societal justice via their purpose statements. This is important because, often, firms superficially claim that they develop stakeholder-led purpose statements (Laufer, 2003) to address the claims of MSs. To counter such superficiality, it is crucial for firms to include MSs from the outset, and as primary stakeholders, so that firms' activities are not only aligned with the interests and expectations of powerful stakeholders but also with those of the MSs and the environment.

The benefits for firms of actively engaging with MSs are multilayered. For example, by being integrative with MSs, firms may no longer need to rely exclusively on the information provided by powerful stakeholders. Firms can examine the demands of the MSs through their MS capabilities approach and allocate resources accordingly. This would also enhance the MSs' trust in firms. If firms are unable to allocate resources according to the expectations of MSs, firms can explain their position directly to MSs. However, firms need to establish a transparent process and structure through which they can perform these activities so that they do not fall into the neocolonial trap of 'business as usual' (Chowdhury, 2021d) but instead transform to become authentically purpose-led. It is important to be aware that informal structures or relationships are not viable in the long run and can disrupt the relationship between the MSs and firms. Firms, thus, need to build a *formal* trust relationship with MSs so that MSs understand the limitations of the firms (McKnight et al., 1998). The development of this type of trust relationship is effective when both parties work together through a common process such as MS capabilities. When firms can build a trust-based relationship with the MSs in this way, they are able to take a more integrative approach in addressing multiple sociopolitical and economic issues simultaneously via their purpose. This is because firms would have integrated mechanisms that do not prioritise one set of stakeholders based on their power and legitimacy attributes (Chowdhury, 2021a).

Again, active engagement with MSs does not mean that the firm should stop sharing information and knowledge with powerful stakeholders, such as government agencies or NGOs. For example, in order to reduce human rights violations, injustice, and inequality, firms, NGOs, and civil society actors need to work together through a cooperative sociostructure (Chowdhury, 2021d). Equally, by actively engaging with MSs, firms may accrue valuable information that can be distributed to government authorities, NGOs, and international organisations, such as the World Trade Organisation and the United Nations, so that these actors can play a coordinating role in developing MS capabilities as a way to ensure MS rights and self-representation. Through this approach, firms can also better serve their internal powerful stakeholders, such as shareholders, because an integrative process that is inclusive and diverse can bring greater sociopolitical and economic success. All the benefits that firms gain from MS capabilities can, ultimately, reduce the level of social misery encountered by MSs, wider society, and the environment and enable the organisation to become truly purpose-led.

Conclusion

My analysis has implications for both firms and managers. Generally speaking, stakeholder management approaches tend to emphasise how managers should engage with their stakeholders (Donaldson & Preston, 1995). However, to date, it is not clear how firms or managers can use such perspectives to create sociopolitical and economic value with and for MSs (cf. Derry, 2012). The problem is that the traditional stakeholder approach is generally embedded within firms' instrumental activities, and this then encourages managers to ignore moral and ethical issues while affording primacy to powerful stakeholders, while overlooking the needs of MSs (Derakhshan & Chowdhury, 2024). Unless firms create sociopolitical and economic value for *all* types of stakeholders, including marginalised groups, then firms will not be able to realise the full potential of their purpose. In other words, if sociopolitical and environmental challenges are not factored in through the inclusion of MSs (since such challenges have the greatest direct and indirect impact on MSs both in the Global South and the Global North), it is likely that an increase in grievances among MSs in various contexts and forms may result in firm-specific disruptions and resistance. This is because MSs are aware of their moral and representational rights and have in-depth knowledge of the importance of their own capabilities in protecting and nurturing the environment where they live or on which they depend.

MSs are also able to utilise different forms of conventional and unconventional mechanisms to achieve their self-representation (Chowdhury, 2021b; Chowdhury et al., 2021). This now becomes much easier, especially in the era of social media. Therefore, firms cannot rely on powerful stakeholders to represent MSs or make use of third parties to solve the problems of MSs. Instead, firms are required to develop organisational purposes and routines that can actualise MS inputs and develop MS capabilities as a pathway to (re)connect and work with MSs for common goals and purposes.

To sum up, MSC provides firms with opportunities to break free from the old way of seeing and doing things with MSs. MSC creates the possibility to challenge the traditional stakeholder approaches and enables firms to discard the rigid routines that currently hinder the development of MS capabilities. More specifically, I conceptualise MS capabilities development as a dynamic process which can revitalise the routine (re)formation of firms. This encourages firms not to consider economic value creation as the core objective; rather, it encourages firms to engage with sociopolitical, economic, and cultural aspects of MS capabilities development that can create the foundation for a purpose that enables both individual freedom and societal well-being. By ensuring individual freedom in and around organisations, firms can reduce their reliance on powerful stakeholders or mitigate the interests of elite actors and communicate and cooperate with MSs. Thus, firms would be able to offer a robust, integrative approach which ensures the representational and natural rights of MSs by fulfilling the dignities that MSs deserve in any interactions and activities in which they (in)directly participate or are affected by.

Further reading

Chowdhury, R. (2023, April 17). Ten years since Rana Plaza: What lessons learned? *Ecotextile Talks*. https://ecotextiletalks.libsyn.com/ten-years-since-rana-plaza-what-lessons-learned

Chowdhury, R., Kourula, A., & Siltaoja, M. (2021). Power of paradox: Grassroots organizations' legitimacy strategies over time. *Business & Society, 60*(2), 420–453. https://doi.org/10.1177/0007650318816954

Editorial Team. (2025, April 24) Reflecting on Rana Plaza. *Eco Age*. https://eco-age.com/reflecting-on-rana-plaza/

Harvard University Press Blog. (2011, May 26). Martha Nussbaum on the capabilities approach to human development. *Harvard University Press Blog*. https://harvardpress.typepad.com/hup_publicity/2011/05/martha-nussbaum-creating-capabilities-human-development.html

Tamvada, J. P., & Chowdhury, R. (2022, August 10). Corporations want to profit from the world's problems—here's how they can solve them instead. *The Conversation*. https://theconversation.com/corporations-want-to-profit-from-the-worlds-problems-heres-how-they-can-solve-them-instead-175284

References

Angell, M. (2005). *The truth about the drug companies: How they deceive us and what to do about it*. Random House.

Batson, C. D., Shaw, L. L., & Oleson, K. C. (1992). *Differentiating affect, mood, and emotion: Toward functionally based conceptual distinctions*. Sage.

Boele, R., Heike, F., & Wheeler, D. (2001a). Shell, Nigeria and the Ogoni. A study in unsustainable development: I. The story of Shell, Nigeria and the Ogoni people—Environment, economy, relationships: Conflict and prospects for resolution. *Sustainable Development, 9*, 74–86. https://doi.org/10.1002/sd.211

Boele, R., Heike, F., & Wheeler, D. (2001b). Shell, Nigeria and the Ogoni. A study in unsustainable development: II. Corporate social responsibility and "stakeholder management" versus a rights-based approach to sustainable development. *Sustainable Development, 9*, 121–135. https://doi.org/10.1002/sd.213

Butler, H. N., & McChesney, F. S. (1999). Why they give at the office: Shareholder welfare and corporate philanthropy in the contractual theory of the corporation. *Cornell Law Review, 84*, 1195.

Chowdhury, R. (2013). *Reconceptualizing the dynamics of the relationship between marginalized stakeholders and multinational firms*. Unpublished doctoral dissertation, University of Cambridge.

Chowdhury, R. (2017a). The Rana Plaza disaster and the complicit behavior of elite NGOs. *Organization, 24*(6), 938–949. https://doi.org/10.1177/1350508416670185

Chowdhury, R. (2017b). Rana Plaza fieldwork and academic anxiety: Some reflections. *Journal of Management Studies*, *54*(7), 1111–1117. https://doi.org/10.1111/joms.12262

Chowdhury, R. (2021a). (In)sensitive violence, development, and the smell of the soil: Strategic decision-making of what? *Human Relations*, *74*(1), 131–152. https://doi.org/10.1177/0018726720955289

Chowdhury, R. (2021b). Self-representation of marginalized groups: A new way of thinking through W. E. B. Du Bois. *Business Ethics Quarterly*, *31*(4), 524–548. https://doi.org/10.1017/beq.2021.16

Chowdhury, R. (2021c). From Black pain to Rhodes must fall: A rejectionist perspective. *Journal of Business Ethics*, *170*(2), 287–311. https://doi.org/10.1007/s10551-019-04183-x

Chowdhury, R. (2021d). The mobilization of noncooperative spaces: Reflections from Rohingya refugee camps. *Journal of Management Studies*, *58*(3), 914–921. https://doi.org/10.1111/joms.12614

Chowdhury, R. (2023). Misrepresentation of marginalized groups: A critique of epistemic neocolonialism. *Journal of Business Ethics*, *186*(3), 553–570. https://doi.org/10.1007/s10551-023-05299-9

Chowdhury, R., Sarasvathy, S. D., & Freeman, R. E. (2024). Toward a theory of marginalized stakeholder-centric entrepreneurship. *Business Ethics Quarterly*, *34*(1), 1–34. https://doi.org/10.1017/beq.2024.1

Chowdhury, R., Kourula, A., & Siltaoja, M. (2021). Power of paradox: Grassroots organizations' legitimacy strategies over time. *Business & Society*, *60*(2), 420–453. https://doi.org/10.1177/0007650320904682

Derakhshan, R., & Chowdhury, R. (2024). Experience of marginalization in noncooperative spaces: The case of undocumented migrant workers in Italy. *Journal of Business Ethics*. https://doi.org/10.1007/s10551-024-05749-1

Derry, R. (2012). Reclaiming marginalized stakeholders. *Journal of Business Ethics*, *111*(3), 253–263. https://doi.org/10.1007/s10551-012-1475-5

Desmond, M. (2016). *Evicted: Poverty and profit in the American city*. Crown Publishing Group.

DiNardo, J. E., & Lee, D. S. (2004). Economic impacts of new unionization on private sector employers: 1984–2001. *Quarterly Journal of Economics*, *119*(4), 1383–1441. https://doi.org/10.1162/0033553042476169

Dodd, E. M. Jr. (1932). For whom are corporate managers trustees? *Harvard Law Review*, *45*, 1145–1163.

Donaldson, T., & Preston, L. E. (1995). The stakeholder theory of the corporation: Concepts, evidence, and implications. *Academy of Management Review*, *20*(1), 65–91. https://doi.org/10.5465/amr.1995.9503271992

Easterbrook, F., & Fischel, D. (1991). The economic structure of corporate law. *Harvard University Press*.

Falguni, A. (2009). *Oil, gas and mineral resources of our country is the blood flown in our vein: Interview with Professor Anu Muhammad*. http://anu-m.blogspot.com/

Freeman, R. E. (1984). *Strategic management: A stakeholder approach*. Pitman.

Friedman, M. (1970, September 13). The social responsibility of business is to increase its profits. *New York Times Magazine*. https://www.nytimes.com/1970/09/13/archives/a-friedman-doctrine-the-social-responsibility-of-business-is-to.html

Gabaldon, T. A. (2006). Like a fish needs a bicycle: Public corporations and their shareholders. *Maryland Law Review*, *65*, 538.

Hacker, J. S. (2006). *The great risk shift: The new economic insecurity and the decline of the American dream*. Oxford University Press.

Henisz, W., Dorobantu, S., & Nartey, L. (2014). Spinning gold: The financial and operational returns to external stakeholder engagement. *Strategic Management Journal*, *35*(12), 1727–1748. https://doi.org/10.1002/smj.2210

King, B. G., & Soule, S. A. (2007). Social movements as extrainstitutional entrepreneurs: The effect of protest on stock price returns. *Administrative Science Quarterly*, *52*(3), 413–442. https://doi.org/10.2189/asqu.52.3.413

Laufer, W. S. (2003). Social accountability and corporate greenwashing. *Journal of Business Ethics*, *43*(3), 253–261. https://doi.org/10.1023/A:1023025018440

Maslow, A. H. (1943). A theory of human motivation. *Psychological Review*, *50*(4), 370–396. https://doi.org/10.1037/h0054346

McKnight, D. H., Cummings, L. L., & Chervany, N. L. (1998). Initial trust formation in new organizational relationships. *Academy of Management Review*, *23*(3), 473–490. https://doi.org/10.5465/amr.1998.926622

Mitchell, R. K., Agle, B. R., & Wood, D. J. (1997). Toward a theory of stakeholder identification and salience: Defining the principle of who and what really counts. *Academy of Management Review*, *22*(4), 853–886. https://doi.org/10.5465/amr.1997.9711022105

Muhammad, A. (2015). Workers' lives, Walmart's pocket: Garments' global chain, from Savar to New York. *Economic and Political Weekly*, *50*(25), 143–150.

Nussbaum, M. (2000). *Women and human development: The capabilities approach*. Cambridge University Press.

Nussbaum, M. (2013). *Political emotions: Why love matters for justice*. The Belknap Press of Harvard University Press.

Pfeffer, J., & Salancik, G. (1978). *The external control of organizations: A resource dependence perspective*. Harper & Row.

Pilkington, E. (2009, June 9). Shell pays out $15.5m over Saro-Wiwa killing. *Guardian*. https://www.theguardian.com/world/2009/jun/08/nigeria-usa

Robeyns, I. (2005). The capability approach: A theoretical survey. *Journal of Human Development*, *6*(1), 93–117. https://doi.org/10.1080/146498805200034266

Sen, A. (1985). Well-being, agency, and freedom: The Dewey lectures 1984. *The Journal of Philosophy*, *82*(4), 169–221. https://doi.org/10.2307/2026132

Sen, A. (1999). *Development as freedom*. Oxford University Press.

Sen, A. (2005). Human rights and capabilities. *Journal of Human Development*, *6*(2), 151–166. https://doi.org/10.1080/14649880500120491

Sheehy, B. (2005). Scrooge the reluctant stakeholder: Theoretical problems in the shareholder-stakeholder debate. *University of Miami Business Law Review*, *4*(Fall/Winter), 193–223.

Shifa, N., Gulrukh, S., & Sumon, M. (2015, May 1). Is the new model of trade unions the answer? *The Daily Star*. https://www.thedailystar.net/op-ed/politics/the-new-model-trade-unions-the-answer-79997

Stevens, B. (1990). Labor unions, employee benefits, and the privatization of the American welfare state. *Journal of Policy History*, *2*(3), 233–260. https://doi.org/10.1017/S0898030600005760

Taher, M. A. (1999). Politicization of trade unions: Issues and challenges in Bangladesh perspective. *Indian Journal of Industrial Relations*, *34*(4), 403–420.

Vidal, J. (2015, January 7). Shell announces £55m payout for Nigeria oil spills. *Guardian*. https://www.theguardian.com/environment/2015/jan/07/shell-announces-55m-payout-for-nigeria-oil-spills

Zyglidopoulos, S. C. (2002). The social and environmental responsibilities of multinationals: Evidence from the Brent Spar case. *Journal of Business Ethics*, *36*(2), 141–151. https://doi.org/10.1023/A:1014411525745

5 Defining and developing organisational purpose

Victoria Hurth and Charles Ebert

Key points

- Organisational purpose is an organisational-level reason to exist that is anchored in serving the good of another, and therefore ultimately in service to long-term well-being for all people and planet—while protecting the basis of long-term well-being for all in the process
- Organisational purpose is also an emotional concept, satisfying the human drive for meaningfulness via a purposeful life through organisational engagement
- Organisational purpose represents a potent but paradigmatically challenging way to re-conceive how we govern the economy, and the organisations that operationalise it
- There is an emerging consensus on what organisational purpose means and how purpose-driven organisations should be governed, as demonstrated by the publication of a national standard in Purpose-Driven Organisations in the UK (PAS 808:2022) and the development of an international standard (ISO) in purpose-driven organisations through the collaboration of 172 national standards bodies
- Defining a purpose and writing a purpose statement should, at a minimum, clarify and justify an organisation's 'reason to exist that is an optimal strategic contribution to long-term well-being for all people and planet' (PAS 808:2022, p. 4) along with committing to not harming the basis of long-term well-being for all in the process
- Defining a purpose, the value it will achieve, and the value that will be protected and regenerated in the process, along with developing a purpose statement to summarise this, can be challenging. It is normally multi-staged and will depend on a number of important considerations

Introduction

An economy is a tool designed by people to optimally allocate and transform resources for collective long-term well-being over time. The definition of sustainable development is in effect the same—we would not have needed the term 'sustainable development' if the economy had been achieving its goal. To 'do good' ultimately

DOI: 10.4324/9781003545224-6

requires alignment with this 'meta-purpose' of long-term well-being for all as otherwise, at some point, harm is inevitable. There are many efforts that, in effect, rethink how we conceive of the market economy as a whole, such as how we measure its success and devise policy for it so that it aligns with collective long-term well-being (e.g. 'Wellbeing Economy' (Fioramonti, 2017), 'Purpose Economy' (Canadian Purpose Economy Project, 2025), or 'Mission Economy' (2021)).

The purpose-driven organisation, as we will outline, is a meso-, organisational-level concept that mirrors these macro-level developments. Both the wellbeing economy and purpose-driven organisations signify a move away from the business-as-usual (BAU) notion that collective long-term well-being is best achieved through an *indirect* focus on financial self-interest. Instead, they anchor their reason for being, and associated governance and strategy, *directly* to the terminal objective of long-term collective well-being—and ensure that strategy in pursuit of this optimising goal also protects and regenerates the resources underpinning long-term well-being in the process. Financial income is understood as the means to this end, rather than an end in itself. As a result, purpose is a challenging yet powerful corrective concept. Whereas businesses have hitherto been told that maximising profits will ultimately drive high levels of collective well-being, in fact, an emphasis on profits as the end goal leads to a misalignment between their decision-making and the kinds of investment and innovation needed to ensure long-term well-being and a sustainable future. In aligning decisions directly with the end goal of the economy, the concept of purpose-driven organisations also helps unify all types of organisations that bring the economy to life. This is because purpose is a concept that is equally applicable to organisations of all kinds, including government as an organisation, corporations (including charities), partnerships, and community groups.

The path to purpose has been necessarily winding, with clarity and consensus often seeming beyond reach. Despite the new era of global consensus-building around the scope and encoding of purpose, which we will cover in this chapter, grasping an overview of organisational purpose as a concept can feel like a daunting task. We hope to somewhat shortcut this journey for you in this chapter.

We (the authors of this chapter) do this on the basis of having immersed ourselves in the complexity of purpose over an extended time period—widely researching the concept, conducting numerous interviews with leading practitioners, and engaging in debates, panels, and roundtables with a diverse range of stakeholders. We have used these insights to write papers (e.g. Ebert & Hurth, 2022; Hurth, 2017; Hurth et al., 2018; Hurth & Stewart, 2022; Hurth & Vrettos, 2021), support the development of executive guidance (CISL, 2023a), present a wide range of lectures, provide corporate and leadership development, and lead purpose-driven standard development (ISO 37000:2021 and PAS 808:2022). From these experiences, we have surmised that people's ability to comprehend the role of purpose today is based on two key factors:

1. How keenly people understand that organisational purpose exists in relation to, and as a deliberate antithesis to, the dominant BAU assumptions that the market

best achieves its goal of collective long-term well-being via orienting companies to their financial self-interest.

2. How far people understand the science of the extreme breakdown of today's social and environmental systems, the corresponding threat to our collective long-term well-being, and the connections between these and our dominant BAU assumptions.

It can be challenging for leaders and academics to comprehend the full extent of the earlier context. The science often seems impenetrable (see Chapters 1 and 2), and the BAU assumptions powerfully shape the world of business and beyond, as we saw in Chapter 4. However, those who understand the urgent scientific imperative for change are best placed to grasp how the context we are facing today gives rise to the need for a fundamental, purpose-led transformation in the societal role of organisations—something that we aim to address in this book.

Those who do not understand this context may nevertheless aspire for their organisations to become purpose-driven, for example, because peers are going down this route, or because of the benefits associated with organisational purpose (e.g. higher levels of talent acquisition and retention, and greater market penetration). These alternative approaches are a start, but they are shallow. They situate purpose as an add-on to existing BAU decision-making, risking the creation of organisational and leadership ambitions that conflict with 'status quo' BAU decision-making and the governance that holds it in place. This undermines the fullest potential of the purpose-driven approach for the organisation. It is because purpose places a response to unsustainability at the centre of the organisation's existence, with financial income as a means to this end (which, in turn, represents a paradigm shift in economic and business thinking) that the concept is so powerful. Hence, purpose cannot be situated simply as a side issue in the hope it will enhance financial income. The potential of purpose to drive positive outcomes via a change to the core of the organisation explains why, despite concerns over 'purpose-washing,' many organisations continue to consider purpose as a mechanism to respond holistically to sustainability challenges—and more so as these challenges intensify.

Because both the theoretical vision and the practical tools are vital, this chapter has two key goals, reflected in its two-part structure. It will begin with a theoretical overview to help leaders avoid a landmine of purpose-washing or stalled efforts resulting from a partial view—reviewing the meaning of organisational purpose and positioning purpose in relation to other macro concepts (e.g. sustainability, values, and ESG (environmental, social, and governance factors) as well as organisation-specific concepts (vision, mission, strategy, and stakeholder engagement)).

In the second section of this chapter, we provide high-level guidance on how to define a purpose and the key components of a purpose statement. This is the starting point for designing a whole governance system that ensures an organisation is purpose-driven. Space constraints mean that we cannot discuss all the details of this governance system here. We would signpost readers instead to The ISO Standard in Purpose-Driven Organisations, ISO37011, and its precursor in the UK, PAS 808. Standards, also known as 'norms,' encode consensus on a topic, revealing views of 'best practice' that exist or that should exist. ISO is an international governance body made up of 172 National Standards Bodies and is the most far-reaching multi-stakeholder forum to develop and encode consensus on standards relating to a topic. Each of the 172 countries has a National Standards Body that develops standards for their country and also proposes, leads, and supports the development of international standards (ISOs). In a signal of conceptual maturity, the first national standard in Purpose-Driven Organisations (BSI PAS 808: 2022) was created by the British Standards Institute (BSI) by an expert representative group via a public consultation process. This process drew on conclusions from the 2021 International Standard in 'Governance of Organisations.' This standard was developed through a five-year process involving 77 countries and 25 official liaisons and signed off by 164 countries, and outlines the foundations of governance with purpose as the primary principle (ISO37000:2021).

The British Standard in Purpose-Driven Organisations outlined the definition of organisational purpose as 'a reason to exist that is an optimal strategic contribution to the long-term well-being of all people and the planet,' along with specifying the associated world views, principles, and behaviours of governing bodies and executive managers of such organisations. We discuss this further in Chapter 10. More recently, a ballot was passed to use this British National Standard to form the working draft in a two-year process aimed at creating the world's first international standard in Purpose-Driven Organisations. The ISO in Purpose-Driven Organisations (ISO 37011) is set to be published in 2026.

The ability to provide this level of robust guidance on purpose-driven organisations is relatively new. It has taken a couple of decades of mainstream interest and experimentation on organisational purpose (supported by far deeper and longer foundations) to reach a level of sufficient clarity to be able to test and encode the consensus surrounding purpose and its potential role in aligning organisations with a sustainable future. This consensus-building and clarification are vital if we are to move from rhetoric to accountable, united action, and help organisations avoid 'purpose-washing.' The path to relative clarity has been long and winding, encapsulating disparate efforts on the part of practitioners and academics to expose and develop aspects of consensus. Part of this journey has been a range of global legal advances away from shareholder primacy and towards enshrining purpose-driven organisations in law, most typically through legislation such as the USA's Benefit Corporation Act and derivatives in countries such as Italy and France (CISL, 2023b). These add purpose alongside assumed shareholder primacy and hence enable, but do not mandate, so-called 'hybrid' or 'dual objective' organisations, rather than promoting fully purpose-driven corporations. Despite these progressions, and sometimes because these laws only partially enable purpose and so obscure the end destination, 'purpose' has traditionally lacked the level of uniform understanding and precision that are required to support universal accountability and measurement.

In order to provide high-level practice guidance in this chapter that anchors to the consensus that is now emerging, we will draw specifically on the two published norms referred to earlier.

What is an organisational purpose?

Definition and background

The term 'purpose' has a substantial history in the organisational literature, with definitions being put forward since the origins of formalised management thinking in the early 1900s (Barnard, 1938; Bradshaw, 1923; Kern, 1919; Tead, 1933). A number of reviews provide insight into the historical foundations and development of the construct (Gartenberg, 2021; George et al., 2021; Hurth et al., 2018; Singleton, 2011, 2014). A deep dive into this history and literature reveals how the roots of organisational purpose as a concept are grounded in human emotional motivation and shared ambition. Early ideas intertwined with more expansive thinking in management at the dawn of the discipline which, during the increasing dominance of BAU thinking from the 1970s onwards, then went through a 'dark age' as human emotions and ethics were placed largely beyond the scope of the firm.

A revival can be observed in the past 15 years as challenges with business-as-usual have led to deeper questions about how business, management, and human behaviour should be perceived. An exposé of this is beyond the scope of this summary chapter. To cut to the heart of the complexity and nuances, we can distil two key dimensions of organisational purpose that tend to lie consistently at the heart of purpose, and which also reveal the nature of the concept as it is popularly used today.

Dimension 1: Organisational purpose is a firm's overarching reason to exist—its overarching objective, identity, and organising principle

Foundationally, organisational purpose is an organising principle that serves to clarify the reason for the organisation's existence both internally and externally. Early definitions of organisational purpose situated it as an overriding aim or objective (Barnard, 1938; Tead, 1933) that served both an internal and external function (Singleton, 2011, 2014). For example, Tannenbaum (1949, p. 226) describes purpose as 'the formal objective of all members of the group.' Clark and Wilson (1961, p. 135) call it the firm's 'explicitly stated substantive goals.' Later definitions also incorporate this aspect. For example, Henderson and Van den Steen (2015, p. 327) describe purpose as 'a concrete goal or objective.' In summary, in one respect, purpose simply is, as per the dictionary definition of purpose, the ultimate objective of the organisation or its 'point.'

What Dimension 1 helps us understand about the relationship between purpose and strategic decision-making

Because purpose is the explicitly stated reason a firm exists, by extension it is also the core component of the organisation's primary governance direction for decisions (Hollensbe et al., 2014), providing 'a guiding light for strategic decision-making' (George et al., 2021, p. 12). In terms of governance, purpose is a whole organisational system concept and therefore is at the centre of the entire governance system, which comprises direction, oversight, and accountability. For example, ISO 37000:2021 has 11 principles, with purpose as the central one that all others orientate around.

George et al. (2021, p. 12) state that being purpose-driven:

> involves the embedding of purpose in organizational routines and practices, the structuring of the organization's business model, partnerships, organizational form, and legal status so that these align with the firm's purpose, and the implementation of governance practices that ensure purpose is an integral part of the firm's management control systems and decision-making.

Purpose, therefore, provides a frame within which coherent, collectively informed decisions can be made and sustained, and a foundation for defending and holding firm on strategic decisions amidst pressure from stakeholders to change course. It also ensures the firm prioritises 'doing good' and achieves positive impact—which, as outlined earlier, will relate either directly or indirectly to well-being outcomes (UN, 2021). The strategic orientating nature of purpose also extends beyond the firm, giving stakeholders tangible insight into what the firm would consider desirable outcomes, enabling pre-emptive action.

What Dimension 1 helps us understand about the relationship between purpose and vision, mission, and strategy

Because organisational purpose is *the* reason for an organisation to exist, along with the parameters that delimit behaviour, it forms the basis of all decision-making. Therefore, high-level scoping concepts like vision, mission, and strategy are necessarily lower-order concepts that are anchored to, and support, the purpose. They serve to make the organisational purpose tangible and provide a level of detail that executives can more easily act upon.

For example, *vision* is fairly consistently understood as being about the details of a desired future state (Collins & Porras, 1991; Ellsworth, 2002; Mirvis et al., 2010). By painting a broad picture of what would be different if the purpose was achieved, internal and external stakeholders can better imagine in more tangible terms what the strategy is trying to realise and the meaningful impacts of the purpose that could arise as a result of achieving the strategic aims. *Mission* describes the highest-level strategic route to achieve the purpose. In summary, purpose is the 'why,' mission is the top-level strategic 'how,' and vision is the description of 'what will be different if we are successful' (Hurth et al., 2018).

To understand how purpose relates to strategy, it is important to set out the overall frame by which organisational decision-making happens, drawing from ISO37000:2021. A governance system structures organisational decision-making through organisational direction, which is then overseen, with accountability maintained through being answerable for performance to significant 'others.' Governance oversight and accountability are therefore anchored to governance direction, which focuses on two key factors: first, the optimising goal of that organisation, its pinnacle reason to exist; and, second, the decision-making parameters (constraints or limitations) that outline how the organisational purpose is to be pursued. Some parameters are imposed on organisations (e.g. the weather, and actions of other actors). Many others are chosen and are ideally encoded in governance policies. Examples include carbon emission quantities, risk tolerance levels, timescales, obeying the law, and organisational values.

Strategy then involves *how*—given the dynamic contexts firms face—the objective can best be achieved (Grant, 2021; Håkansson & Snehota, 1989) *within* the decision-making parameters (Carver, 2006; ISO37000). The high-level answer to the 'how' questions just outlined is a term that has already been described: *mission.* Through further strategic elaboration about how the governing body intends for the mission to be achieved, for example, via strategic objectives and targets, the organisation executive creates and delivers strategy that cascades throughout the organisation as it seeks to achieve the purpose. Purpose thus provides the basis for strategy, vision, and mission.

Dimension 2: Purpose as the pursuit of the 'good' and the route to meaningfulness and sustainability

The second defining aspect of an organisational purpose is that the organisation's reason to exist is anchored to serving the good of others, as opposed to the self-interest that is the basis of traditional business expectations and governance. The evidence shows that the contemporary concept of organisational purpose is not the same as the dictionary definition of 'the point of something,' regardless of the *nature* of its point. Instead, organisational purpose moves beyond describing the functional activities of a company or product-level ambitions, and into the realm of providing the basis for core human meaningfulness and motivation (Bhattacharya et al., 2023; Collier, 1998a; Gartenberg, 2021; Hollensbe et al., 2014).

At the individual level, 'people all desire and need a motivating force beyond self-interest' (Irving, 2018, p. 27). Purpose 'is a deep and abiding sense of transcendent meaning that connects the individual to a larger story.' More specifically, 'purpose ignites our innate "telic orientation"... and sets the individual on a course of energetic striving toward noble ends' (McCloskey, 2014, p. 90). In other words, purpose is acknowledged as a fundamental human motivation for a life that feels worthwhile, because that life is judged to make someone else's better.

An organisation is a collective of individuals who seek to act 'as one,' and as a human organism, it is therefore reflective of the human condition. Work provides a pathway through which individuals can live out meaningful lives having the potential to fulfil 'how one makes a difference through service to others and, in doing so, [how one] derives meaning and purpose in life' (Fry, 2003, p. 703).

Given the earlier, it is not surprising that individual purpose at the micro level is connected to purpose at the organisational level, with early thinkers describing organisational purpose as relating to a spiritual state (Kern, 1919, p. 445) or a source of morale (Bradshaw, 1923, p. 494). Bartlett and Ghoshal (1994, p. 88) describe it as 'a company's moral response to its broadly defined responsibilities,' with similar conceptualisations in works by Bhattacharya et al. (2023), Collier (1998), and Rindova and Martins (2023). As a result, organisational purpose 'tends to emerge from a broader set of societal values and expectations and captures a higher order purpose that links to moral and ethical obligations' (George et al., 2021 p. 2; see also Collier, 1998). This also means that purpose relies on the firm's understanding of its potential 'harms' in order to establish what, in contrast, 'good' outcomes mean.

The connection between organisational purpose and 'doing good' anchors purpose-driven companies to establish a purpose that is based on the assessment of the benefits and harms to the long-term well-being of others, relevant to the time period. In the past, the notion of optimising well-being arose in a different context. For example, Unilever made easy-to-use soap products like Lifebuoy in the UK as a way to 'make cleanliness commonplace' in an era of industrial slums (Unilever, 2025, p. 1). Today, while many of these historical issues are still relevant, the context of harm and benefit to collective long-term well-being is very different. It is now far more apparent that

- All life on Earth is interconnected and the breakdown in environmental and social systems threatens the collective well-being of all life, forever
- In the vast majority of cases, societally harmful effects arise from organisational decisions
- The concept of well-being is the clearest and most enduring expression of the ultimate value we are trying to create and not to destroy (for human and non-human life)
- The economy is the main tool we use to allocate resources to achieve this collective long-term well-being
- The definition of sustainability reflects the mandate of governments and the definition of the goal of the economy

For example, the connection between the ability to achieve our collective well-being ('development'), the role of the economy in achieving this, and the fact that it is organisations in the economy that operationalise the transformation of resources for collective long-term well-being, has long been known. The very definition of sustainable development and sustainability is anchored to the Brundtland report, a consequence of international leaders coming together to call for a new way to enact the economy (World Commission on Environment and Development, 1987) and achieve collective well-being over the long term. To achieve a change in the economy requires a change in organisational decisions, including (but not only) in businesses—which, in turn, requires a change in the governance systems that structure routine decision-making in any organisation. The anchoring of an organisation's existence to directly achieve well-being outcomes is commonly found in conceptualisations of purpose (Collier, 1998; Ferrarini, 2020; Hollensbe et al.,

2014; Mayer, 2021; Zahra et al., 2009) and reflected in practice (e.g. Ebert et al., 2018; Radjou & Prabhu, 2015).

Governments enable organisations, including businesses, to exist and govern how this happens so that they achieve their mandate. It is clear that governments exist to deliver well-being for those within their remit. As the International Public Sector Accounting Standards Board (2024) states bluntly: 'the primary function of governments and other public sector entities is to provide services that enhance or maintain the well-being of citizens and other eligible residents.' The economy is a tool designed by society and utilised by government to achieve this mandate, hence, by extension, the economy exists to maximise the well-being of the collective over time. Or, put another way, the role of the economy is about: 'allocating the resources available to society in a way that maximizes social wellbeing' (Goodland & Ledec, 1987, p. 20), with economics defined as: 'a study of mankind in the ordinary business of life; it examines that part of individual and social action which is most closely connected with the attainment and with the use of material requisites of wellbeing' (Marshall, 1890, p. 2).

Hence, the organisations that operationalise the economy are, via the economy, society's key mechanism that takes resources and transforms them for collective long-term well-being, so that the goal of sustainability and the goal of governments and the economy are aligned. This is also resonant of wider notions of purpose. As Drucker (1954) noted,

> if we want to know what a business is we have to start with its *purpose*. And its purpose must lie outside of the business itself. In fact, it must lie in society since a business enterprise is an organ of society.

The earlier, along with the fact that collective notions of 'good' and 'value' are ultimately anchored to contributing long-term well-being for all, means that this notion of 'long-term well-being for all' could be viewed as the closest expression of humanity's overall optimising goal or 'meta-purpose.'

Organisational purpose and its relationship to other concepts

These two dimensions provide the foundation for the consensus-based definition in the British Standard in Purpose-Driven Organisations where Organisational Purpose is defined as: '[a] reason to exist that is an optimal strategic contribution to the long-term wellbeing of all people and planet' (p. 4).

How organisational purpose relates to 'business-as-usual' organisations

The deeply established economic view, formalised by Friedman and the 'Chicago Boys' and embedded during the post-war era in global legal and cultural systems, asserts that the executive exists as a servant of the shareholder (Friedman, 1970). This, along with a series of assumptions about how a market works to automatically maximise for collective well-being over time, created the moralised basis for financial self-interest at the economic, organisational, and individual levels to be accepted and embedded in governance systems as a distant proxy for well-being outcomes (Diesendorf et al., 2024; Marquis, 2024; Mayer, 2018; Monbiot & Hutchison, 2024). Organisational purpose, in anchoring the reason for an organisation to exist to *directly* serve the long-term well-being of others, stands in stark contrast to this notion of shareholder primacy (see also Chapter 4).

More recently, Larry Fink, CEO of BlackRock, the world's largest financial asset manager, expressed similar sentiments in his famous letter to his shareholders:

> society increasingly is turning to the private sector and asking that companies respond to broader societal challenges. Indeed, the public expectations of your company have never been greater. Society is demanding that companies, both public and private, serve a social purpose.
>
> (Larry Fink, 2018)

Mark Carney, the then governor of the Bank of England, also noted 'true purpose, including that of business, must be grounded in the objectives of clients, society and humanity' (Civil Service Quarterly, 2019). These public statements reflect shifting legal and governance agenda towards enabling purpose-driven organisations to pursue well-being outcomes directly, rather than via financial self-interest (see Hurth, 2023; Mocanu & Ventura, 2023).

If a purpose-driven organisation is to contribute to long-term well-being for all, it also logically requires the avoidance of harms in achieving this goal. As such, purpose not only requires the organisation's reason to exist (the overarching innovation agenda) to be justified in terms of long-term well-being for all, but at the same time, the organisation is logically required to ensure that it creates and adheres to organisational parameters while devising strategy to achieve its aims. In other words, to ensure that the thresholds of healthy social and environmental systems are assured when strategic decisions are made. Any less, and they might actually be creating more harm in the process, even if the company may be doing some good in other areas.

Organisational purpose and organisational values

In the management literature, Collins and Porras (1991, p. 35) define organisational values as 'the organization's basic precepts about what is important in both business and life, how business should be conducted, its view of humanity, its role in society, the way the world works, and what is to be held inviolate.' More basically, values are core to world views, which are beliefs about what is most valuable to protect and enhance in the world: 'values . . . underlie our worldviews' (Mifsud & Sammut, 2023, p. 2). Hence, values are very broad and foundational—for example, innovation, fairness, honesty, and creativity. Organisational purpose and organisational values are therefore intricately connected—organisational values inform the organisational purpose and values also form key parameters for how the purpose is achieved (ISO 37000: 2021). For example, a firm's purpose may be eliminating malaria in Ghana, and the firm's world views and values regarding the importance of stopping treatable curtailments to life will have informed the choice of this purpose. At the same time, treating people fairly (equitable pay) and respecting fundamental human rights (not relying on slavery in the supply chain) would be additional organisational values that, whilst not the reason the organisation exists, are values about how to treat people that should be adhered to while achieving the organisational purpose.

The extent to which an organisational purpose aligns with company values influences how strongly or easily the purpose resonates with the company and its employees (Ebert, under review). Ellsworth (2002, p. 334) notes the important relationship between organisational purpose and employee values, and argues that management has the role of motivating and engaging employees by helping them see how their values are 'grounded in a worthy corporate purpose.'

Organisational purpose and ESG

A few accounts of the history of ESG now exist (Pollman, 2022; Hardyment, 2024). ESG began as a voluntary mechanism through which investors who valued more than just financial returns could pressure companies to reduce the harms caused to the well-being of others by corporate actions (often taking the form of specific exclusion criteria, such as tobacco or arms). Today, ESG has become a fast-moving political process aimed at embedding broadly held societal concerns about issues such as climate change, biodiversity loss, and social injustice into corporate decision-making processes. Ultimately, society is seeking verification that companies are not asset-stripping shared resources and thereby undermining our collective well-being. ESG then can be viewed as a 'do no harm' agenda that seeks, initially through reporting, to bring organisational business models within thresholds of healthy social and environmental systems.

There is therefore a key relationship—and a clear difference—between organisational purpose and ESG. Organisational purpose is concerned with achieving an optimal strategic contribution to the goal of long-term well-being for all, and ESG is about not doing harm to the basis of long-term well-being for all while achieving that goal. Because organisations that are purpose-driven are anchored to the 'meta-purpose' of long-term well-being for all, it makes logical sense that they would be advanced in driving forward the ESG agenda to ensure the meta-goal is not harmed. Consequently, the most commonly used definition of organisational purpose contains reference both to the optimising goal (the purpose itself) as well as to the parameters that need respecting in the process: 'to produce profitable solutions to the problems of people and planet,' and 'not to profit from producing problems for people or planet' (Mayer, 2021, p. 889). 'Profitable' indicates that, naturally, any activities need to ensure the adequate stocks and flows of financial capital for the organisation to be able to remain viable and to achieve its goal. This definition was popularised as part of the British Academy's *Purpose of the Corporation* programme, led by Professor Colin Mayer of the University of Oxford. Mayer was also an expert contributor to the development of the PAS 808 and this definition underpinned the development of definition of purpose used in the standard. We explore how purpose may be measured further in Chapter 10.

Organisational purpose and CSR/CSV

The executives we have interviewed over the years have nearly always considered corporate social responsibility (CSR) to be an add-on—an important device for keeping up a public image but not central to the business, and an activity that gets abandoned when budgets are tight (Hurth et al., 2018). Academic definitions are more varied but tend to hold the same general view of CSR as being an 'add-on' (see Freeman & Liedtka, 1991). Some have attempted to address this problem of adoption through what is known as 'broad CSR' (Schwartz & Saiia, 2012), which has the potential to integrate people and planetary concerns into a company's core operations (Cheng et al., 2014). Purpose differs from broad CSR in that it *forms* the core operations rather than being integrated into it.

In a review of 37 different definitions of CSR put forward in the academic literature, Dahlsrud (2008) notes an absence of clarity over how to evaluate the performance of CSR activities: 'the definitions do not provide any descriptions of the optimal performance or how these impacts (the categories of responsibility or impact) should be balanced against each other in decision-making' (Dahlsrud, 2008, p. 6). Therefore, while both CSR and organisational purpose can include a wide variety of societal concerns and well-being outcomes, organisational purpose involves deciding on a specific commitment as the firm's reason for being and embedding additional concerns, such as protecting social and environmental systems and how people should be treated, into decision-making parameters. CSR, on the other hand, even in the broadest sense, involves balancing multiple obligations in a more general way—usually as a side agenda to core operations.

Another related term is creating shared value (CSV). In their seminal paper, Kramer and Porter (2011, p. 6) describe CSV as 'policies and operating practices that enhance the competitiveness of a company while simultaneously advancing the economic and social conditions in the communities in which it operates.' In other words, CSV promotes the classic approach of seeking 'win-wins'—here, specifically, between financial income (which remains the pinnacle objective) and well-being outcomes of varying levels and expressions which are optimised via profit maximisation. What it does not do, however, is challenge the notion that financial income maximisation should be the core optimising goal. Hence, where win-wins are not possible and trade-offs are required, profit maximisation will remain the logical priority and leading focus for strategy and investment cases. Organisational purpose, on the other hand, necessarily situates profits as a means to an end—a vital capital required to pay the bills, to meet the expectations of stakeholders, and to invest in achieving the organisation's purpose. Purpose-driven organisations will also naturally want to seek win-wins so that the purpose can be maximised and the healthy functioning of social and environmental systems are not just protected but further enhanced in the process. Therefore, insights from CSV are also likely to be of use for purpose-driven organisations.

In summary, the term organisational purpose has a long tradition in history and practice. Contemporary notions of organisational purpose have two fundamental dimensions. First, purpose is a firm's overarching reason to exist that establishes the firm's ultimate value creation objective and is the core focus of strategy and all other decisions. Second, purpose is the pursuit of the good of another and is increasingly understood as the firm's specific and optimal contribution to long-term well-being for all. This makes organisational purpose unique, although related to various concepts, such as ESG, CSR, CSV, and organisational values. Finally, because purpose is the highest priority for the firm, all these other factors, including financial performance, become a means to that end, factors whose healthy functioning needs to be protected through behavioural parameters established (and then overseen and accounted for) in the governance system.

Becoming a purpose-driven organisation

Defining your organisational purpose

Since the explosion of interest in organisational purpose, a wide range of consultants globally have tackled the important, and often lucrative, question of how a company should think about the foundational step of defining an organisational purpose. Deciding on an appropriate purpose can be problematic. In particular, it can be challenging to define a purpose that is scoped and elaborated at the right level to provide enough strategic direction while not stifling strategic agility. The key questions that need to be asked and answered when deciding a purpose include:

1) Is the well-being outcome sufficiently ambitious and impactful in relation to long-term well-being for all?
2) Is it specific enough to serve as an anchor for strategic decision-making?
3) Is it broad enough not to stifle strategic creativity and innovation?
4) Are you sure this is *the* overarching objective for which you exist, rather than actually an important *means* to an end, for example, ensuring your organisation operates within the appropriate legal boundaries, or within a chosen boundary such as the net-zero emission limit?
5) Who else is pursuing this well-being outcome and are they able to do it better? Could you work together, or should you instead focus your efforts on a different organisational purpose?
6) What are the assets, talents, stakeholders, and other resources that you will need, and can you access them?
7) How can we measure, assess, and hence oversee and account for progress towards it?

While varying advice exists on the precise aspects to consider when defining an organisational purpose, ISO 37000 and PAS 808 contain guidance that shortcuts useful steps. The key guidance is replicated in Table 5.1 along with supporting reflection questions. As you will note—many of these are adjusted components of a

Table 5.1 ISO37000 factors for determining an organisational purpose and reflection questions.

ISO 37000 and PAS808 guidance on aspects to consider	*Reflection questions*
The problems or opportunities for long-term well-being for all	
'important issues including global threats that evolve over time (e.g. climate change)' (ISO 37000) 'the most urgent and important problems faced in respect to long-term wellbeing for all people and planet, and the material factors influencing this (e.g. by utilizing a materiality assessment)' (PAS 808) 'future trends affecting . . . the nature of the problems facing long-term wellbeing for all' (PAS 808)	• What are all the major issues facing long-term well-being for all and what drives them? • How are they likely to evolve over time? • What are others doing about these or likely to be doing?
The organisational values and identity	
'the organizational purpose, and the parameters within which the organizational purpose is to be pursued, are determined on the basis of a clearly defined set of organizational values' (ISO 37000)	• What are the organisation's existing values? • How aligned are these with the world views that purpose-driven organisations are likely to hold? • What are the aspirational world views of the organisation? • Based on the existing and aspirational world views, what are the most important things the organisation should seek to protect and enhance?
'existing documentation relating to the organizational purpose and the scope of the organization's activities, such as constituting documents or other artefacts' (ISO 37000)	• What are the key written materials that can help the organisation understand what it currently commits to achieving, for example, founder speeches, previous legal constitutions, and earlier promotional material?
'the historic, current and aspirational core identity of the organization, including the organizational values and expected ethical behaviour' (ISO 37000)	• What were the firm's original intentions? Did it have an organisational purpose or an intention to address important issues for others when it was first constituted? • In which markets and market offerings has the firm traditionally been active? • What do the internal and external stakeholders think the firm's central activity is? What would they like it to be? • Where does the firm have authority and loyalty? Where does it have a social license to operate?

(*Continued*)

Table 5.1 (Continued)

ISO 37000 and PAS808 guidance on aspects to consider	*Reflection questions*
Stakeholder influence	
'stakeholder expectations' (ISO 37000)	• What are the world views and values of those that the firm affects, or who affect the organisation (see later for a stakeholder relationship classification)? How would these change if the purpose was one option or another, and how would this support or constrain the ability of the organisation to achieve that purpose?
Organisational and market context	
'organizational capabilities and opportunities' (ISO 37000)	• What are the firm's resources and capabilities? What are those areas of expertise relevant to?
'the organization's core capabilities or potential capabilities to address the problems identified' (ISO 37000) 'what other organizations are already doing to solve the problems identified, and particularly where other organizations might be better placed to do so' (PAS 808) 'future trends affecting the organizational context' (PAS 808)	• What is the firm in a position to impact? What societal good is it well-positioned to improve? What societal good is it well-positioned to improve that others are not as well-positioned to improve?

classic 'environmental scan' that any organisation would undertake before making a strategic decision about how they will achieve a goal—here, the stràtegic decision is: 'What can the organisation do (as its reason to exists) that can best achieve long-term wellbeing for all?'

This list reflects both an 'inside out' and 'outside in' approach to identifying the well-being outcomes that form the basis of an organisational purpose (Hurth & Vrettos, 2021a). There will be a range of supporting actions within these areas of analysis that can be useful for both identifying the organisational purpose and then enacting it. One of the most important is developing a clear conceptualisation of stakeholders (we discussed stakeholders in more depth in Chapter 4). This is a vital step in clarifying the nature of the relationships that will be created by virtue of different candidates for the purpose—and therefore what information, engagement, and influence over decision-making will be created, or are required. Following this analysis for each purpose option, it should become clear if the stakeholder constellation that results is aligned or where it may be misaligned.

It is then important to consider how adaptable these stakeholders are and whether therefore one purpose is more feasible than another. Here, the British Standard in

Purpose-Driven Organisations (British Standards Institute, 2022, p. 22) offers a clear, purpose-specific stakeholder taxonomy to work with:

- **Ultimate beneficiary stakeholders**: all people and planet over the longer term, including future generations
- **Primary (or focal) beneficiary stakeholders**: those whose well-being the organisational purpose primarily serves. In the case where the organisational purpose is focused on addressing an ultimate social or environmental system problem (e.g. climate change or inequality), the ultimate beneficiary stakeholders would be all people and planet
- **Influential stakeholders**: those the governing body is legally bound to (e.g. member stakeholders) and those it chooses to account to (e.g. reference stakeholders)[1]
- **Enabling stakeholders**: those who the organisation relies on to achieve the organisational purpose in the way intended
- **Affected stakeholders**: those who are affected by the decisions the organisation makes, even if their effect on the organisation is non-existent, unperceivable, or very limited

If there is not enough clarity or certainty regarding the chosen organisational purpose, or if key powerful stakeholders who could block the process from the outset have been identified, then it is worth considering a 'pre-purpose phase' (Hurth & Vrettos, 2021b) during which an action plan to create the right conditions is designed and enacted.

Developing your organisational purpose statement

Purpose statement crafting

An organisational purpose and the value it seeks to achieve, along with the key values that will be protected, need to be made concrete and explained in an organisational purpose statement, and then further elaborated upon as appropriate for relevant internal and external stakeholders so that it can be used to guide decisions effectively and represent the organisational identity. Often, an organisational purpose statement is interpreted as being the short strapline that is communicated internally and externally, for example, Anglian Water: 'Love Every Drop.' However, this is the top-line communication tool. Behind this should sit a detailed statement that sets out, justifies, and scopes the organisational purpose to give enough strategic guidance and to set the value intent that the governing body is accountable for the organisation achieving. The firm's legal constitution should then be adjusted to reflect this at a top-line level, ensuring it contains nothing that would restrict achievement of the purpose. The British Standard in Purpose-Driven Organisation (2022, p. 22) outlines that the organisational purpose statement forms the 'highest-level strategic scoping of an organization' and clarifies 'the specific contribution it (the organization) will make towards its general goal of serving the long-term wellbeing of all people and planet.'

The standard states that the organisational purpose statement needs to include three components:

- **Component 1, Detail sufficiency:** An organisational purpose statement should include enough detail about the desired results the organisation wants to create so that it is evident to internal and external stakeholders that the purpose is the optimal contribution to long-term well-being for all people and planet (the meta-purpose). Doing this stage effectively will require a well-developed 'theory of change'; plotting the assumed cause and effects together provides evidence-based links between the organisational purpose and the meta-purpose.
- **Component 2, Non-negotiable commitment to no harm:** Reflecting Mayer's (2021) definition of organisational purpose, the purpose statement should also outline a non-negotiable commitment 'to not harming, not benefiting from harm, and where possible increasing the health of social and environmental systems . . . in the process of achieving the purpose' (British Standards Institute, 2022, p. 22). Specifically, this means ensuring that the organisation achieves, at a minimum, a fair and adequate contribution to the healthy functioning of the social and environmental systems that underpin the long-term well-being of all people and the planet.

  This would require an understanding of the full suite of social and environmental issues relevant for long-term well-being for all, and the issues threatening this healthy functioning, such as flood risks, extreme heat, infectious diseases, food security, biodiversity loss, inequality, and loneliness (we discussed many of these in earlier chapters of this book). These would then be used to estimate the organisational behaviour that would result in a fair and adequate contribution to maintaining or enhancing (regenerating) social and environmental functioning. For example, a level of carbon emissions per year that is in line with a trajectory for zero carbon emissions by a certain date; paying people at least a living wage; ensuring water drawdown of less than a certain level. Sources such as the Sustainable Development Goals (SDGs), Planetary Boundaries, and the book *Doughnut Economics* (Raworth, 2018) outline key thresholds for minimum conditions of healthy functioning of social and environmental systems.
- **Component 3, Organisational parameters:** To govern for the commitment to no harm would also require setting decision-making parameters that ensure that this commitment is respected by decision-makers, overseen and able to be accounted for. As PAS 808 (p. 22) outlines, the purpose statement 'should include the highest-level (*organizational*) parameters for the organization to follow e.g. red lines/floors/guardrails while the organization is achieving the purpose.' Together with the purpose, these parameters comprise the governance direction to which the organisational decision-making, governing system, and governing body for accountability are anchored. Parameters are the limiting constraints to be respected by all decision-makers while optimising the purpose. Where decision-making parameters cannot be respected immediately, interim

parameters are needed to clarify the path and timescale to this point—and any reparation needed. The core decision-making parameters that keep strategy across the organisation aligned with healthy social and environmental systems are foundational. Also foundational are those parameters that ensure achievement of the minimum value intended for the organisation's stakeholders—either because they require this value or, informed by organisational values, the organisation wants to achieve this minimum threshold value. This can include establishing average minimum levels of financial return to shareholders or ensuring suppliers are treated fairly. It is also vital to create parameters that ensure that oversight and accountability (the two other governance functions alongside direction) happen effectively.

- **Component 4, Strategic organisational parameters:** As well as core parameters, there are others that the governing body can strategically decide to create and include in the purpose statement—because they are judged to be important to ensure decision-making achieves the purpose within the other parameters. These include, for example, risk limits, decision-making processes, or leadership styles. Key to this is ensuring that organisational values are lived. The organisational values, which derive from organisational world views about what is most valuable to protect and enhance and how the world works, will have been created as part of the process to decide the organisational purpose. These values are also important in establishing the behaviours it will adhere to when carrying out all its activities—and so will inform all parameters but may require others. One example might be 'treating people with care' through an associated decision-making parameter of 'no bullying behaviour' and 'paying people a minimum wage throughout the supply chain.'

Testing the organisational purpose and purpose statement

Once the organisational purpose has been decided and the purpose statement developed, it is vital to evaluate their usefulness within the organisational decision-making context. This could be through a 'beta testing' phase where the company works with the organisational purpose while actively eliciting the views of employees on how meaningful, ambitious, and useful it is in directing firm-wide decisions, and those of relevant stakeholders. For established organisations, this phase will also better reveal how far the governance system and stakeholder constellation need to be adjusted so that the purpose can be achieved.

One of the most important elements of the organisational purpose statement is testing its level of scope. Many organisations scope their organisational purpose far too broadly so that, in effect, it simply repeats the meta-purpose and defines its values as a list of universal ethical approaches any organisation could hold (e.g. 'make everyone's lives better, forever'). Instead, it is important to try to ensure that the organisational purpose and the fully specified statement explaining and justifying it provides enough detail for it to serve as a useful and unifying guide for day-to-day decisions, strategy development, and assessment. At the same time, it needs to be sufficiently broad to allow strategy-makers across the firm to apply it to their contexts effectively, creatively allowing as much room for dynamic agility as possible.

Elaborating the purpose statement

While establishing the right level of scope for the purpose is vital, wherever the line is drawn, the purpose can be further elaborated if needed to provide more specific guidance. This guidance allows for adaptability as it is open to more frequent change than the purpose statement itself.

As PAS 808 stipulates, this elaboration should include a 'stakeholder clarification statement' to guide decision-makers as to how different stakeholders should be taken into account when making organisational decisions, using the five-part classification described earlier. This enables decision-makers to consider and involve different types of stakeholders in clear and consistent ways that are aligned with the purpose and with the expectations set with different stakeholders (see also Chapters 4 and 7).

Crafting the organisational purpose statement and elaborating its implications are critical steps in the firm's new purpose-driven journey as they establish the governance direction. This direction then needs to be embedded in governance policy and cascaded across the organisation through layers of strategy making and management policy—direction that is then overseen and accounted for. For established organisations, this will almost always require a whole organisational governance review and adjustment as the previous governance system will be optimised to achieve a different goal and within different parameters.

Oversight mechanisms should include continual monitoring of the relevance of the purpose statement. Decision-makers and stakeholders should be able to raise issues where decisions to maximise the purpose within the parameters fall short of aligning with the meta-purpose of long-term well-being for all. It is vital not to get stuck with a purpose that is not 'fit for purpose.'

The role of a purpose summary or strapline

A pithy summary or even shorter strapline of the purpose can be a vital communication mechanism for spreading awareness of the purpose and for providing a quick mental access point for the purpose statement detail. This may need to be adjusted for different stakeholders but ideally would be a singular jumping-off point for everyone. One common problem is viewing this purpose strapline as the focus of the entire purpose development process, rather than simply seeing it as a communication tool used at the end of a process which focuses on establishing a justifiable, meaningful reason to exist that guides decision-making. (We explore purpose communication further in Chapter 9).

These, sometimes hollow, statements are unfortunately one reason why some say they have lost faith in the concept of organisational purpose (Findlay & Moran, 2019). Even though the purpose slogan should be easily communicable, it should be rich enough to hold meaning. For example, the South African financial services firm Old

Mutual summarises its purpose as: *to help our customers thrive by enabling them to achieve their lifetime financial goals, while investing their funds in ways that will create a positive future for them, their families, their communities and broader society.*

Although it may not align with 'long-term well-being of all' as it aims to maximise financial returns without necessarily considering the broader impacts on the well-being of the investor, society, or the planet over time from these returns, the core 'parameters' detailed (*while investing their funds, etc.*) are aligned with preserving the basis of long-term well-being for all. What is clear is that this statement serves as a summary that provides enough strategic depth to aid stakeholders in their decision-making. This allows them to engage with the summary as a jumping-off point, without needing to constantly refer back to the complete purpose statement for any real sense of decision-making guidance.

Concluding thoughts

Interest in organisational purpose has been growing at the same time as the realisation that our global economy is driving a dangerously unsustainable world. There is increasing recognition that placing financial self-interest as the problem to be solved in organisations, and financial capital as the main resource to be protected, is taking humanity way off track with respect to the goal of long-term well-being for all that the economy purports to achieve. Purpose is the first organisational concept that removes financial self-interest as the organising principle and replaces it with direct strategic endeavour towards our shared value creation goal—long-term well-being for all. By aligning routine decision-making with a sustainable future via an explicitly designed governance system, purpose has the possibility to pivot a highly ineffective and inefficient economy (one that is destroying rather than achieving its ultimate goal) into a highly efficient and effective one. Achieving sufficient stocks and flows of financial capital (along with all other types of capital) becomes one necessary strategic problem to solve as a *means* to achieving the purpose—not the point of decision-making. Although the concept of organisational purpose has necessarily travelled the path of experimentation, fragmentation, and potential co-option by the existing system (purpose-washing), we are entering a new phase. With the establishment of international standards that codify the emerging consensus and create a shared language and accountability frame, the future looks promising.

While it has not been possible to outline the full governance system of a purpose-driven organisation in this chapter, PAS 808 outlines key aspects and can be freely accessed. There is an implementors' group on LinkedIn ('PAS 808: purpose-driven organisations') that is free to join and provides an opportunity to learn from peers on their implementation journeys. PAS 808 was a leap forward in establishing clarity and consensus about what it means to be a purpose-driven organisation, however, the bigger step to an international consensus is vital and under way. When published (forecast to be 2026), the ISO in Purpose-Driven Organisations (ISO 37011) will launch a new phase in the journey of organisational purpose and its role in aligning the world with a sustainable future.

Note

1 **Reference stakeholder**

stakeholder to whom the governing body has decided to account when making decisions pertaining to the organisational purpose

Note 1: For example, scientific advisory board to a research organisation, parents of the pupils in a school, community advisory boards for companies.

Note 2: In some cases, a member stakeholder can also be a reference stakeholder.

Further reading

Ebert, C., & Hurth, V. (2022). Principles of purposeful business: illustrative examples. *Journal of the British Academy*, *10*(5), 163–207.

Hurth, V., Renshaw, B., & Fioramonti, L. (2025—forthcoming). *Beyond profit: Purpose-driven leadership for a wellbeing economy*. John Murray Business. https://doi.org/10.5871/jba/0010s5.163

ISO 37000: 2021 Governance of Organizations.

PAS 808. (2022). *Purpose-driven organizations—worldviews, principles and behaviours for delivering sustainability—guide*. Edition 1. British Standards Institute. https://www.iso.org/standard/65036.html

References

Anglian Water. (2023). *Our purpose*. Anglian Water Services. https://www.anglianwater.co.uk/corporate/about-us/our-purpose/

Barnard, C. I. (1938). *The functions of the executive*. Harvard University Press.

Bartlett, C. A., & Ghoshal, S. (1994). Changing the role of top management: Beyond strategy to purpose. *Harvard Business Review*, *72*(6), 79–88.

Bhattacharya, C. B., Sen, S., Edinger-Schons, L. M., & Neureiter, M. (2023). Corporate purpose and employee sustainability behaviors. *Journal of Business Ethics*, *183*(4), 963–981. https://doi.org/10.1007/s10551-022-05090-5

Bradshaw, F. F. (1923). Companionable books. *Journal of Social Forces*, *1*, 492–494. https://doi.org/10.2307/3004973

British Academy. (2018). *Reforming business for the 21st century: A framework for the future of the corporation*.

British Standards Institute. (2022). *PAS 808*. https://www.bsigroup.com/en-GB/standards/pas-808/

Canadian Purpose Economy Project. (2025). https://purposeeconomy.ca/

Carver, J. (2006). *Boards that make a difference: A new design for leadership in nonprofit and public organizations* (Vol. 6). John Wiley & Sons.

Cheng, B., Ioannou, I., & Serafeim, G. (2014). Corporate social responsibility and access to finance. *Strategic Management Journal*, *35*(1), 1–23. https://doi.org/10.1002/smj.2131

CISL (University of Cambridge Institute for Sustainability Leadership). (2023a). *CISL's business transformation framework: Preliminary diagnostic*. Cambridge Institute for Sustainability Leadership. https://www.cisl.cam.ac.uk/download-business-transformation-framework-preliminary-diagnostic

CISL (University of Cambridge Institute for Sustainability Leadership). (2023b). *Future of boards legal and regulatory frameworks for sustainability* (Phase 1, Part 2). https://www.cisl.cam.ac.uk/download-future-boards-research-study

Civil Service Quarterly. (2019). In conversation with . . . Mark Carney, Governor of the Bank of England. *Civil Service Quarterly*. https://quarterly.blog.gov.uk/2019/02/13/in-conversation-with-mark-carney-governor-of-the-bank-of-england/

Clark, P. B., & Wilson, J. Q. (1961). Incentive systems: A theory of organizations. *Administrative Science Quarterly*, *6*(2), 129–166. https://doi.org/10.2307/2390752

Collier, J. (1998). Theorising the ethical organization. *Business Ethics Quarterly*, *8*(4), 621–654. https://doi.org/10.2307/3857544
Collins, J. C., & Porras, J. I. (1991). Organizational vision and visionary organizations. *California Management Review*, *34*(1), 30–52. https://doi.org/10.2307/41166682
Dahlsrud, A. (2008). How corporate social responsibility is defined: An analysis of 37 definitions. *Corporate Social Responsibility and Environmental Management*, *15*(1), 1–13. https://doi.org/10.1002/csr.132
Diesendorf, M., Davies, G., Wiedmann, T., Spangenberg, J. H., & Hail, S. (2024). Sustainability scientists' critique of neoclassical economics. *Global Sustainability*, *7*, e33. https://doi.org/10.1017/sus.2024.36
Drucker, P. F. (1954). *The practice of management: A study of the most important function in American society*. Harper and Brothers.
Ebert, C., & Hurth, V. (2022). Principles of purposeful business: Illustrative examples. *Journal of the British Academy*, *10*(s5), 163–207. https://doi.org/10.5871/jba/0010s5.163
Ebert, C., Hurth, V., & Prabhu, J. (2018). *The what, why, and how of purpose: A guide for leaders*. https://www.managers.org.uk/~/media/Files/
Ellsworth, R. R. (2002). *Leading with purpose: The new corporate realities*. Stanford University Press.
Ferrarini, G. (2020). An alternative view of corporate purpose: Colin Mayer on prosperity. *Rivista delle società*. https://doi.org/10.2139/ssrn.3552156
Findlay, S., & Moran, M. (2019). Purpose-washing of impact investing funds: Motivations, occurrence and prevention. *Social Responsibility Journal*, *15*(7), 853–873. https://doi.org/10.1108/SRJ-11-2017-0260
Fink, L. (2018). Larry Fink's 2018 letter to CEOs: A sense of purpose. *BlackRock*. https://www.blackrock.com/corporate/investor-relations/2018-larry-fink-ceo-letter
Fioramonti, L. (2017). *Wellbeing economy: Success in a world without growth*. MacMillan.
Freeman, R. E., & Liedtka, J. (1991). Corporate social responsibility: A critical approach. *Business Horizons*, *34*(4), 92–105. https://link.gale.com/apps/doc/A11015279/AONE?u=anon~8b8e465b&sid=googleScholar&xid=3f1ed1a2
Friedman, M. (1970, September 13). The social responsibility of business is to increase its profits. *The New York Times*, 17.
Fry, L. W. (2003). Toward a theory of spiritual leadership. *The leadership quarterly*, *14*(6), 693–727. https://doi:10.1016/j.leaqua.2003.09.001
Gartenberg, C. M. (2021). *Purpose-driven companies and sustainability*. University of Pennsylvania. https://doi.org/10.2139/ssrn.3786823
George, G., Haas, M. R., McGahan, A. M., Schillebeeck, S. J. D., & Tracey, P. (2021). Purpose in the for-profit firm: A review and framework for management research. *Journal of Management*, *49*(6), 1841–1869. https://doi.org/10.1177/01492063211006450
Goodland, R., & Ledec, G. (1987). Neoclassical economics and principles of sustainable development. *Ecological Modelling*, *38*(1–2), 19–46. https://doi.org/10.1016/0304-3800(87)90043-3
Grant, R. M. (2021). *Contemporary strategy analysis*. John Wiley & Sons.
Håkansson, H., & Snehota, I. (1989). No business is an island: The network concept of business strategy. *Scandinavian Journal of Management*, *5*(3), 187–200. https://doi.org/10.1016/0956-5221(89)90026-2
Hardyment, R. (2024). *Measuring good business: Making sense of environmental, social and governance (ESG) data*. Routledge. https://doi.org/10.4324/9781003457732
Henderson, R., & Van den Steen, E. (2015). Why do firms have "purpose"? The firm's role as a carrier of identity and reputation. *American Economic Review*, *105*(5), 326–330. https://doi.org/10.1257/aer.p20151072
Hollensbe, E., Wookey, C., Hickey, L., George, G., & Nichols, C. V. (2014). Organizations with purpose. *Academy of Management Journal*, *57*(5), 1227–1234. https://doi.org/10.5465/amj.2014.4005
Hurth, V. (2017). Organizations as open systems that need purpose and integrated thinking. *Board Leadership*, *2017*(150), 1–8. https://doi.org/10.1002/bl.30069

Hurth, V. (2023). *Unleashing governance for sustainable business*. University of Cambridge Institute for Sustainability Leadership. https://www.cisl.cam.ac.uk/files/unleashing_governance_paper.pdf

Hurth, V., Ebert, C., & Prabhu, J. (2018). *Organisational purpose: The construct and its antecedents and consequences*. Working Paper No. 02/2018, Cambridge Judge Business School, University of Cambridge. https://www.jbs.cam.ac.uk/wp-content/uploads/2020/08/wp1802.pdf

Hurth, V., & Stewart, I. S. (2022). Re-purposing universities: The path to purpose. *Frontiers in Sustainability*, *2*. Article 762271. https://doi.org/10.3389/frsus.2021.762271

Hurth, V., & Vrettos, A. (2021a). *Unleashing the sustainable business: How purposeful organisations can break free of business-as-usual*. University of Cambridge Institute for Sustainability Leadership. https://www.cisl.cam.ac.uk/resources/unleashing-sustainable-business

Hurth, V., & Vrettos, A. (2021b). *Unleashing the sustainable business. Paper 2: How purpose transforms an organisation*. University of Cambridge Institute for Sustainability Leadership.

International Organization for Standardization (ISO). (2021). *ISO 37000:2021 Governance of organizations*. https://www.iso.org/standard/65036.html

International Public Sector Accounting Standards Board. (2024). *Handbook of international public sector accounting pronouncements* (Vol. 1). International Federation of Accountants. https://www.ipsasb.org/publications/2024-handbook-international-public-sector-accounting-pronouncements

Irving, J. A. (2018). Leader purposefulness and servant leadership. In D. Van Dierendonck and K. Patterson (Eds.), *Practicing servant leadership* (pp. 25–42). Springer International Publishing. https://doi.org/10.1007/978-3-319-75644-8

Kern, F. (1919). Recht und Verfassung im Mittelalter. *Historische Zeitschrift*, *120*(JG), 1–79.

Kramer, M. R., & Porter, M. (2011). *Creating shared value* (Vol. 17). FSG. https://environz.wordpress.com/wp-content/uploads/2012/11/creating-shared-value-porter-and-kramer-harvard-business-review-article.pdf

Marquis, C. (2024). *The profiteers: How business privatizes profits and socializes costs*. PublicAffairs.

Marshall, A. (1890). *Principles of economics*. Palgrave Macmillan.

Mayer, C. (2018). *Prosperity: Better business makes the greater good*. Oxford University Press.

Mayer, C. (2021). The future of the corporation and the economics of purpose. *Journal of Management Studies*, *58*(3), 887–901. https://doi.org/10.1111/joms.12660

McCloskey, M. (2014). *Learning leadership in a changing world: Virtue and effective leadership in the 21st century*. Springer. https://doi.org/10.1057/9781137476371

Mifsud, R., & Sammut, G. (2023). Worldviews and the role of social values that underlie them. *PLoS One*, *18*(7), e0288451. https://pmc.ncbi.nlm.nih.gov/articles/PMC10370735/

Mirvis, P., Googins, B., & Kinnicutt, S. (2010). Vision, mission, values: Guideposts to sustainability. *Organizational Dynamics*, *39*(4), 316–324. https://doi.org/10.1016/j.orgdyn.2010.07.006

Mocanu, D., & Ventura, L. (2023). *Future of boards: Legal and regulatory frameworks for sustainability (Phase 1: Part 2)*. https://www.cisl.cam.ac.uk/download-future-boards-research-study

Monbiot, G., & Hutchison, P. (2024). *Invisible doctrine: The secret history of neoliberalism*. Penguin.

Pollman, E. (2022). *The making and meaning of ESG*. University of Pennsylvania Institute for Law and Economics, Research Paper No. 22–23. SSRN. https://papers.ssrn.com/sol3/papers.cfm?abstract_id=4219857

Radjou, N., & Prabhu, J. (2015, February 10). Frugal innovation: How to do more with less. *The Economist*. https://amzn.eu/d/8GSQAkw

Raworth, K. (2018). *Doughnut economics: Seven ways to think like a 21st century economist*. Chelsea Green Publishing.
Rindova, V. P., & Martins, L. L. (2023). Moral imagination, the collective desirable, and strategic purpose. *Strategy Science*, *8*(2), 170–81. https://doi.org/10.1287/stsc.2023.0190
Schwartz, M. S., & Saiia, D. (2012). Should firms go 'beyond profits'? Milton Friedman versus broad CSR. *Business and Society Review*, *117*(1), 1–31. https://doi.org/10.1111/j.1467-8594.2011.00397.x
Singleton, L. (2011). *Understanding the evolution of theoretical constructs in organization studies: Examining cooperation and purpose*. Boston College. http://hdl.handle.net/2345/2582
Singleton, L. (2014). Understanding the evolution of theoretical constructs in organization studies: Examining 'purpose'. In *Academy of Management Proceedings* (p. 14367). Academy of Management Briarcliff Manor. https://doi.org/10.5465/ambpp.2014.246
Tannenbaum, R. (1949). The manager concept: A rational synthesis. *The Journal of Business of the University of Chicago*, *22*(4), 225–241. http://www.jstor.org/stable/2349754
Tead, O. (1933). *Human nature and management: The applications of psychology to executive leadership*. McGraw-Hill.
Unilever. (2025). *Lifebuoy*. https://www.unilever.co.uk/brands/personal-care/lifebuoy/
United Nations (UN). (2021). *SDG impact standards: Enterprises*. https://sdgprivatefinance.undp.org/sites/default/files/resource-documents/SDG-Impact-Standards-for-Enterprises-Version1-EN.pdf
World Commission on Environment and Development. (1987). *Our common future*. Oxford University Press.
Zahra, S. A., Gedajlovic, E., Neubaum, D. O., & Shulman, J. M. (2009). A typology of social entrepreneurs: Motives, search processes and ethical challenges. *Journal of Business Venturing*, *24*(5), 519–532. https://doi.org/10.1016/j.jbusvent.2008.04.007

6 Effecting the change to become purpose-led

Craig V. Robinson and Luis D. Torres[1]

Key points

- Organisational change can be messy, and a purpose-driven transformation is likely to be large in scale and complex in nature
- Shifting to a purpose-driven approach requires an assessment of the organisation's current state, engagement with stakeholders, and empowering leaders and employees
- Resistance to change always arises in some form; with the right approach, leaders seeking a purpose-driven transformation can use resistance to gather feedback, engage with local issues, and better embed purpose
- Embedding reporting and accountability mechanisms, including ESG and purpose-driven KPIs, can help to drive sustained commitment and trust from employees, managers, and external stakeholders over time

Purpose-driven organisational transformation

In today's complex global economy, firms face challenges that go beyond economic profitability. Traditional business models that prioritise profit maximisation at the expense of social and environmental impacts are less socially acceptable in a world that is interconnected, and understands (despite some ongoing debate) the need for more sustainable approaches to the environment as well as the social value of work. As we have seen throughout this book, the concept of being 'purpose-driven' has emerged as a response to this evolution, encouraging organisations to pursue social and environmental ambitions alongside economic ones.

This shift has led to a revised set of expectations for leaders. In 2019, nearly 200 CEOs of major United States' companies, and members of the Business Roundtable, redefined the purpose of a corporation, shifting focus from exclusively maximising shareholder value to serving all stakeholders. In 2024, the Business Roundtable published a review of the progress five years on, underlining the fact that increased value is created for long-term shareholders by supporting employees and customers, serving as good partners to suppliers, and enhancing communities.

DOI: 10.4324/9781003545224-7

The key message is twofold. First, that business leaders need to have a clear idea of what their organisation's wider purpose is and that a purpose-driven approach does actually create economic value.

For any organisation, becoming purpose-driven requires persistent effort that goes beyond simply developing a positive purpose statement. A purpose must feel true, be embedded in the culture of the organisation, and be rooted in the organisation's core activities. Purpose-driven transformation is about shifting the entire operating model, culture, and strategy to align them to the purpose. Therefore, being a purpose-driven organisation is a radical change to the business-as-usual model that requires a structural disruption in leadership to enact this change. Further, such change involves engagement with employees at every level. A purpose-driven plan of engagement can be helpful in helping employees shape the meaning of their work more explicitly.

Embedding and delivering purpose-driven outcomes remains a challenge. Where a gap arises between what leaders say their purpose is, and what society is seeing, organisations that appear to be 'purpose-washing' and making misleading promises risk being punished by the media, by customers, and in the stock market. This chapter, grounded in ideas related to organisational change management and strategy implementation, is intended to provide a practical guide to managing the process of purpose-driven transformation. We start with a discussion of the challenges of becoming purpose-driven and then move on to a set of recommendations to support leaders in successfully leading and delivering a purpose-driven transformation.

The case for purpose-driven transformation

Despite the advantages of adopting a purpose, it can be difficult for leaders to do so. One of the most important challenges concerns shifting business leaders' mindset away from a focus on maximising short-term profit (Lyon et al., 2024). Business leaders have deeply rooted beliefs regarding profits and shareholder primacy that impede thinking of alternative approaches. Stock market volatility and a 24-hour media cycle further encourage a focus on short-term changes and profitability. This creates a tension and requires careful navigation and good leadership, and is particularly an issue for large corporations where there are strong financial incentives for managers and investors to focus only on profit maximisation (see also Chapter 5).

Leaders who commit to embedding purpose in their organisations need to carefully navigate this challenge. In some cases, leaders may prioritise purpose-driven initiatives that lack a clear path to financial health. If such initiatives fail to show how they improve the bottom line, leaders risk alienating investors and the board of directors. Box 6.1 shows an example of the risks that purpose-driven CEOs face when investors focus primarily on profit maximisation.

Box 6.1 The purpose-driven leader's dilemma

Emmanuel Faber, former CEO of Danone—a global producer of yogurts such as Activia and Oikos, and Evian water, among others—turned the company into an 'entreprise à mission' in June 2020. This is a legal category created in France to recognise companies whose purpose is far broader than profits and growth. Faber named his strategy 'One Planet, One Health,' led a push into healthy foods, and created a carbon-adjusted earnings per share indicator, measuring Danone's success directly in relation to its environmental performance. However, the company's sales and margin growth had underperformed compared with some competitors in recent years and worsened during the pandemic. In March 2021, Faber would step down as chairman and CEO, ejected by the board of directors after the company faced growing calls from some shareholders to replace him. Danone's shares jumped in early trading following the news of Faber's departure. What does this all mean for purpose-driven leaders? As ever, CEOs face an impossible dilemma: maximise profits to please shareholders in the short term but ignore the longer-term social and environmental impacts of corporate actions; or embed purpose thereby potentially anger shareholders who may face short-term reductions in their returns due to the investments needed to turn purpose into a reality. Faber had placed these issues at the core of Danone, but shareholders were unable to see the benefit.

Source: Adapted from Walt (2021)

Society's expectations concerning business pursuing a purpose beyond profit maximisation have been typically incorporated by business in the form of corporate social responsibility (CSR) initiatives. As we saw in Chapter 5, CSR is not the same as purpose. CSR can be understood as the 'policies and practices of corporations that reflect business responsibility for some of the wider societal good' (Matten & Moon, 2008, p. 405). CSR would then be anything that a business does to respond to society's expectations beyond the minimum legal requirements. This sounds similar to purpose-related initiatives, but CSR is frequently implemented as an 'add-on' or 'side dish' to the core business. Because the actual goals for most CSR efforts are not directly connected to corporate functions (e.g. marketing, sales, and operations), many companies are not certain what goal CSR should accomplish or what strategies to implement. The result is often a collection of unfocused, unlinked, and unrelated initiatives that fail to align with the core business. In an attempt to meet this challenge, industry associations, civil society organisations, and other international organisations have responded by proposing hundreds of CSR standards and tools, with the influx at times leading to confusion among managers, CEOs, and investors (Torres et al., 2023).

Running a purpose-driven business is frequently viewed as being synonymous with sacrificing financial returns. Our current financial system works based on short-term profit-seeking strategies and a neglect of the indirect negative effects of production (negative externalities). In this context, any strategy that does not provide a quick win or attempts to internalise negative externalities is seen as a failure. This misconception has shifted the debate to the concept of sustainable finance and its broad operationalisation as environmental, social, and governance (ESG) issues. The ESG agenda received a renewed commitment after the 2008 financial crisis and then the Paris Agreement in 2015 (see Box 6.2). Both concepts—CSR and ESG—are related, as they attempt to promote greater responsible business conduct. The scope is, however, different. While CSR focuses on what socially responsible actions companies take, ESG focuses on the way they measure sustainability performance and publicly disclose it. Without a well-implemented purpose, ESG metrics and reporting mechanisms suffer from a lack of legitimacy, trust, and business alignment. We discuss this issue further in Chapter 10.

Box 6.2 A renewed commitment to sustainable finance

Financial crises reduce trust in businesses. This is what happened after the 2008 Financial Crisis. A strong consensus emerged from policymakers, regulators, civil society, and financial services organisations themselves, that many financial institutions had engaged in harmful activities through the ruthless pursuit of short-term profit without considering their positive social purpose. Indeed, the outcome of the 2008 Financial Crisis had negative impacts both economically and socially across the world. The financial sector needed to fundamentally reconsider its strategies, operations, and activities, and align these with more socially purposeful, longer-term objectives. The answer was that finance should become more sustainable and less focused on short-term gains, reshaping the activities and operations of banks, asset managers, insurers, stock exchanges, investors, and other financial institutions to consider the broader economic, environmental, and social impacts of their work.

In recent years, the emphasis has shifted to how finance can support sustainable economies and societies more generally, including the natural environment. Financing sustainable economic, environmental, and social objectives, often those set out in the UN Sustainable Development Goals, has become a fundamental objective. More recently, and particularly from 2015 with the signing of the Paris Agreement on Climate Change, policymakers, regulators, and financial services practitioners have focused on environmental sustainability in particular. This has included significantly increasing the financing of new technologies and activities designed to reduce greenhouse gas emissions (climate change mitigation) and/or support climate-resilient development (climate change adaptation), as well as identifying and disclosing climate-related financial risks.

Yet, acknowledging these issues across what are often large, multinational, and multifunctional financial organisations is neither easy nor straightforward. Banks have had to reorganise, reprioritise, and refocus their objectives, performance metrics, and organisational cultures in order to adapt and succeed in the new reality, and making such changes requires a clear and structured approach.

The challenges and societal shifts outlined in this section should make the need for purpose-driven change in organisations clear. The question then becomes—how to do it and how to do it successfully?

Changing to become purpose-driven

Becoming purpose-driven aims to change the way things are done within organisations and there is a vast literature on organisational change that suggests it is both complex and requires some thought. Corporate purpose needs to be materialised, made intelligible, and be observable and measurable inside and outside the organisation (Aguilera, 2023). Being purpose-driven is more than simply a statement about the reason an organisation exists and why it is a force for social good. Alongside an articulation of purpose, a tangible shift in how an organisation operates is vital.

The remainder of this chapter is devoted to considering the steps required to implement a purpose-driven change successfully, but before moving into the different activities required to deliver change, it is worth considering how change in organisations works. Burnes (2017) proposes that the nature of change in an organisation depends on two different parameters:

- The speed of change, because the urgency of change required depends to some extent on the situation the organisation finds itself in. Large-scale transformations are more likely to happen in turbulent environments than in those that are relatively stable
- The scope of change, because more drastic and urgent challenges likely require a greater degree of change than minor, more emergent issues. Broader change programmes are likely to encompass the entirety of the organisation, while smaller-scale programmes will focus on particular tasks, processes, attitudes, or behaviours and will target particular groups or individuals

The scope of change required to reorient an organisation towards a particular purpose is likely to be broad, involving change across the whole organisation; an incremental, continuous improvement approach will not take the organisation far enough to become purpose-driven. Moving an organisation from a traditional model to a purpose-based model is not just about improving what they do or a gradual adjustment to existing strategy statements. This is why a purpose-led approach goes beyond the CSR and ESG approach, which incorporates social and environmental concerns into an organisation's work in an evolutionary manner without radically changing their overall intent or scope of operations.

In short, becoming purpose-driven entails a radical challenge to business-as-usual. Radical change requires a shift in the fundamental values of the organisation and the way it operates (Todd, 1999). The time period over which this occurs, however, will be driven by both internal and external factors.

Externally, environmental disruption or crises (as discussed in Box 6.2) can act as triggers for rapid organisational change. Internally, changes in senior leadership often precipitate changes in strategy or, indeed, a reorientation of the organisation's purpose.

The development of a purpose statement is covered in detail in Chapter 5; the remainder of this chapter is focused on what needs to happen next to embed purpose into the DNA of the organisations. Organisations must ensure that their purpose resonates with their key stakeholders and is aligned to their core values and long-term goals. Such alignment can then act as the compass that directs all decisions. In the absence of such coherence, even a well-designed and carefully thought-out purpose risks becoming disconnected from the organisation's reality. The following sections cover six different recommendations related to organisational change that leaders need to take account of when embedding purpose in their organisation.

Identify your starting point

The process of embedding organisational purpose is not straightforward. The first key question any leader needs to ask is: How prepared is my organisation to become purpose-driven? The response should provide insight into current business priorities and stakeholder attitudes to social good, including social and environmental challenges.

The following five-stage model may help answer this question (adapted from Torres, 2017). Each stage is a description of the organisation's current approach to value creation, its relationship with stakeholders, and its view on the organisation's environmental and social role. The model gives an indication of the organisation's current capacity to implement a transformation, taking into account present organisational priorities, the level of commitment of key stakeholders, the ability and capacity of key individuals (change agents or allies), and the extent to which internal and external stakeholders perceive a purpose-driven shift as a necessary move. The five stages are not necessarily sequential; some organisations may exhibit characteristics of different stages at the same time. Nevertheless, it provides a useful starting point for assessing readiness for change, with organisations at later stages providing a more fertile ground for a purpose-driven change programme than those at earlier stages.

The University of Cambridge Institute for Sustainability Leadership (CISL) provides a set of diagnostic tools that purpose-driven leaders can use to assess their organisation's level of readiness to undertake a purpose-driven transformation. These focus on four areas of activity as follows (CISL, 2023):

1. Align business purpose, strategy, and business models
2. Set evidence-based targets, measure, and be transparent
3. Embed sustainability in practice and decisions
4. Engage, collaborate, and advocate change

Box 6.3 The five stages of organisational preparedness

Stage 1: ***Protective.*** Purpose beyond profit maximisation is not part of the organisation's interest. Any activities with a social and environmental focus will aim at managing risks and improving/maintaining reputation.

Stage 2: ***Compliance-seeking.*** The organisation focuses on compliance with the regulatory framework, while also striving to meet minimum CSR/ESG industry standards. A basic understanding of what purpose is starts to emerge as long as regulations or CSR/ESG standards explicitly mention it.

Stage 3: ***Capability-seeking.*** The business rationale for working towards social and environmental issues begins to develop. A more systematic approach to stakeholder engagement is in place, and top management is committed to contribute to broader goals. Purpose is still seen as a social responsibility commitment parallel to what the business is and does.

Stage 4: ***Value-seeking.*** Top management becomes sensitised to the potential for long-term improvements, business-wide opportunities, and the benefits of a purpose-driven organisation. Purpose is understood to contribute to long-term viability and success of the organisation.

Stage 5: ***Transforming.*** The organisation has undergone a profound change in its culture and strategy. Business leaders adopt fully transparent practices and aim to disseminate the benefits of a purpose-driven organisation within and across industries. Existing collaborations with community groups, other companies, NGOs, and the government transform into durable alliances aiming at realising the purpose.

Each of these areas of activity can be assessed according to how purpose-focused they are on a spectrum from short-term self-interest (least ready) to purpose-driven (fully embedded). A fifth cross-cutting theme, individual and collective leadership, assesses the foundational leadership principles, mindsets, and practices that underpin the actions of individual and collective agents across the business. We turn to the leadership issue shortly, but first explore ideas concerned with resistance to change.

Consider how to approach resistance

The potential for resistance to change to arise during organisational change initiatives is often cited as a primary reason for their failure (Kotter, 1995). Resistance can emerge for several reasons, broadly reflecting the reactions of individuals or groups when they perceive that a change threatens their current state. Resistance is best understood as a negative response to the process or outcomes of organisational change, or more broadly as any conduct that serves to maintain the status quo in the face of pressure to alter it. Clearly, when implementing a purpose-driven transformation, resistance may arise, but with the right approach and an awareness of the nature and types of resistance to change, it can be leveraged to provide useful feedback and provide further input into the early-stage diagnostic work.

Traditional approaches treat resistance to change as an individual dispositional trait grounded in behaviours such as routine seeking, emotional reaction, short-term focus, and cognitive rigidity (Oreg, 2003). Seminal work by Piderit (2000) framed resistance as a tripartite issue relating to behavioural, cognitive, and emotional reactions to change. A key distinction for leaders to be aware of is the difference between *active* and *passive* resistance, where an individual or a group may voice concerns and resist openly, or alternatively go along with things they do not believe in while remaining silent and withdrawing emotionally. Both forms of resistance present a problem for leaders seeking to embark on a purpose-led transformation.

If a change programme is framed and communicated carefully, however, resistance to change can be useful. Given that the scale of a purpose-driven transformation is likely to be large, the resistance that emerges presents an opportunity to engage in dialogue and gather feedback on what may work best in different parts of the organisation. This is in line with more progressive theories on resistance to change, which frame resistance as an inevitable consequence of human and organisational life rather than as malicious behaviour. Indeed, resistance can be a source of innovation and reflection (Thomas & Hardy, 2011) and leaders may inadvertently exacerbate resistance through poor planning and communication (Ford et al., 2008). In short, an open dialogue-based approach to change is more likely to surface resistance and provide a useful mechanism for feedback in a purpose-driven transformation. Engaging employees early in the process through discussions, feedback sessions, and workshops that allow them to contribute ideas and voice concerns is likely to surface useful feedback and create a sense of ownership. This requires a particular approach to change leadership, which we address next.

Empower leadership at all levels

We examine leadership in the context of decision-making in Chapter 7; in this section, we explore the role of leaders across the organisation in delivering change. As a precondition for any successful purpose-driven transformation, senior leadership and management of the organisation, including the board, must make an authentic, firm, and public commitment to purpose. A major hindrance to the shift to being purpose-led is a lack of support from the board, and research suggests that when they are held accountable for their social and environmental actions, they are more likely to take purpose seriously and include it in their oversight activities (Carballo, 2023). Without the support of the board, CEOs are likely to find themselves pushing an agenda that will be lost as soon as they leave (Lyon et al., 2024).

While a purpose-driven transformation may be initiated by senior leaders, those in leadership roles across the organisation have a critical role to play in making the transformation successful. Middle leaders, situated between senior leadership and frontline employees, are well-placed to act as change agents who drive, facilitate, and sustain change within organisations (Balogun & Johnson, 2004). This intermediary position allows middle managers to translate strategic visions into practice, manage ambiguity, and mobilise support from below. In a purpose-driven transformation, they are critical in mediating between top-down directives and local realities, translating organisational purpose into a series of locally relevant actions. In order to do this, however, they need licence to operate and to innovate; their capacity to succeed as change agents depends on the support they receive from more senior leadership and the cultivation of trust and influence both upwards and downwards (Huy, 2001).

Middle leaders often maintain daily contact with staff and can draw upon informal networks and peer influence to promote change (Currie & Procter, 2005). Their embeddedness in the organisational culture gives them credibility and an empathetic understanding of employee concerns. While this closeness can create tension, a successful purpose-driven transformation needs middle leaders to reconcile their loyalty to their own team with loyalty to upper management (McConville, 2006). Research shows that midlevel support for the purpose of the business and clarity in the path towards that purpose result in better financial performance (Gartenberg et al., 2019).

Engage key stakeholders

As discussed in detail in Chapter 4, a true purpose-driven transformation must consider the interests of all stakeholders. The direct or indirect engagement of all the stakeholders allows the organisation to properly understand their expectations. Customers seek companies that align with their values (Kramer & Porter, 2011), employees want to work for organisations that stand for something meaningful to them (Gioia et al., 2013), and investors are increasingly interested in ESG factors when making decisions (Eccles & Klimenko, 2019). Thus, purpose must resonate not only internally among employees but also externally with customers, investors, local communities, suppliers, non-governmental organisations, and even policymakers (Andriof et al., 2017).

There exists, however, a tension between the objectives and interests of different stakeholders, for example, between investors and the wider community. Resolving this tension requires authentic leadership and the balancing of shareholder and stakeholder interests (Freeman et al., 2007). This entails a process of stakeholder identification, mapping, and engagement. Typically, identification follows standard categories (government, clients, etc.), and these are repeated across organisations and sectors (Greenwood, 2007). Stakeholder salience may vary, but organisations need to avoid overlooking marginal voices (Reed et al., 2009). Employee consultation can uncover additional relevant stakeholders based on operational knowledge (Bryson, 2004) and external groups such as NGOs can provide insights into community needs and marginalised populations (Ebrahim, 2003). An inclusive approach to stakeholder engagement strengthens the authenticity of the organisation's purpose and builds goodwill across diverse communities. These stakeholders should also include any group or individual experiencing any kind of deprivation such as those affected by poverty, inequality, and disability. Effective purpose-driven organisations should not overlook the differences between distinct groups of people as well as the disparities between different classes, racial groups, regional populations, and so on.

Once stakeholders are identified, the next step is active listening. Purpose-driven organisations must focus on understanding the needs, expectations, and values of their stakeholders. This involves open channels for communication, such as surveys, focus groups, social media interactions, and town hall meetings (Noland & Phillips, 2010). It is essential to approach these engagements by treating stakeholders as equal partners rather than as passive recipients of organisational decisions on purpose. Listening fosters trust, lays the groundwork for collaboration, and ensures that the company's purpose aligns with the values of those it serves (Senge et al., 1999).

Not all stakeholders have the same needs, expectations, or concerns. For instance, investors may prioritise financial returns and ESG metrics, while employees may value a clear alignment between their work and the company's purpose. Similarly, not all stakeholders hold equal weight in shaping or responding to a company's purpose-driven initiatives. Stakeholder mapping helps categorise stakeholders based on their influence, interest, and proximity to the company's activities. By doing so, purpose-driven companies can focus their resources and efforts on those stakeholders who are most invested in their mission or have the power to significantly impact outcomes. Stakeholder mapping also uncovers potential risks or points of resistance (Bryson, 2004; Ackerman & Eden, 2011). For example, understanding opposition from specific groups allows companies to address concerns proactively and adapt their strategies to mitigate conflict (Post et al., 2002).

The final step is to create a plan of engagement. By engaging stakeholders in co-creating initiatives, organisations can harness diverse perspectives, innovative solutions, and shared enthusiasm (Prahalad & Ramaswamy, 2004). Regular feedback loops allow organisations to refine their purpose-driven strategies while maintaining alignment with stakeholder priorities. This dynamic engagement fosters resilience, builds mutual trust, and supports the organisation's societal legitimacy (Heugens et al., 2002).

Embed purpose in your culture

If engagement with stakeholders and empowered leadership are important, then so is the organisation's culture when it comes to leading change. Culture can be summarised very simply as 'the way things are done around here' (Drennan, 1992) and consists of the values, beliefs, and behaviours that organisational members share (Schein, 2010). These shared beliefs are not necessarily written down, however, and evolve socially through interactions, the symbols that people associate with success, the power structures that exist, and the behaviours that are rewarded (Johnson, 1992). The fact that culture is pervasive, enduring, and implicit (Groysberg et al.,

2018) makes it challenging to manage, and leadership commitment is central to fostering a culture that prioritises purpose (Quinn & Thakor, 2018). As discussed in the preceding sections, leaders need to create an environment in which employees feel empowered to contribute to the mission and take pride in their roles within the organisation. When leaders nurture a culture of purpose, they inspire their teams to internalise the company's values and become active participants in its transformation (May et al., 2004).

Companies may say they are committed to the purpose, but without a supportive culture, employees will not be empowered to enact these shared values in their work. Purpose-driven transformation that does not engage with the organisation's culture is likely to result in only cosmetic changes, akin to following a 'ticking the box' approach to keep up appearances. Organisations can superficially adopt internal management strictures or manipulate them to 'appear' compliant with a purpose without 'being' compliant by changing the corporate culture (Torres, 2024).

Corporate purpose is closely related to culture as a construct stemming from the shared meanings, motivations, and practices of people within the organisation that is projected externally through the organisation's reputation for contributing to social ends (Almandoz, 2023). Purpose-driven cultures can foster a sense of belonging and motivation among employees, as they enable employees to perceive their work as contributing to a greater cause rather than merely serving financial goals. Employees who identify with their company's purpose exhibit higher levels of motivation, engagement, and satisfaction (Parmar et al., 2019; van Tuin et al., 2020). Therefore, the motivating power of corporate purpose ultimately depends on whether individuals and teams experience it as personally meaningful (Almandoz, 2023).

Myriad frameworks are available for diagnosing cultural issues and setting about how to change them; one likely to help with a purpose-driven transformation comprises the four actions proposed by Groysberg and colleagues (2018), which we include as a suggested further reading at the end of this chapter for those interested. In short, they suggest first articulating the ambition in terms of the type of culture change required—for example, in moving from an individualistic and competitive culture to a more collaborative one. They also suggest seeking out, developing, and promoting leaders who exemplify the target culture and using this to facilitate conversations about the importance of culture change through conversations and dialogue. Finally, they suggest underpinning the new culture with changes to organisational design, part of which involves the measuring and reporting of performance, to which we turn next.

Measure and report performance

It should be clear from the preceding discussion that an organisation's purpose, even once defined, is not likely to be universally agreed on at all levels. The priorities of stakeholders differ and a commitment to environmentally friendly operations or social equity may be praised by one group but challenged by another; existing

Box 6.4 Cisco's approach to culture and purpose

Cisco's Conscious Culture is a foundational element of its organisational identity and aligns very clearly with its overarching purpose, which the company articulates as Powering an Inclusive Future for All.

Francine Katsoudas, Chief People, Policy & Purpose Officer at Cisco, explains the culture as follows:

> *Let's demonstrate empathy. Let's be kind to one another. Let's have curiosity. Let's continue to learn. And most importantly, let's take action. In a Conscious Culture, that's our job. We take action if we experience something or if we see something. And in doing so, we change the culture for all.*

Cisco's culture is built around six principles designed to guide daily decision-making and interactions:

- Give Your Best—encourages colleagues to strive for excellence, share knowledge, and embrace diverse ideas
- Give Something of Yourself—promotes authenticity and trust in interactions with colleagues, customers, and wider stakeholders
- Give Your Ego the Day Off—encourages prioritising collaborative solutions over individual recognition and supporting the best ideas regardless of origin
- Take Accountability—empowers individuals to take ownership of their actions and decisions
- Take Difference to Heart—recognises that varied perspectives strengthen the collective
- Take a Bold Step—encourages innovation and courage in pursuing new ideas, even when they challenge the status quo

Cisco's Conscious Culture is designed to help integrate the company's purpose into its operations and is managed through three different areas. First, the environment, where Cisco aspires to foster a workplace where everyone feels welcome and celebrated. Second, they leverage their unique characteristics as an organisation. Cisco's principles are deeply embedded in its ethos, often serving as a compelling reason for individuals to join and remain with Cisco over a long period. This links to the third and final aspect, the Cisco Experience, which focuses on how individuals, teams, and leaders interact to do meaningful work.

Cisco was named the world's best place to work by Forbes in 2019 and has consistently been recognised as a leading purpose-driven organisation with a deeply embedded culture. They also have a purpose reporting hub on their website, which outlines the organisation's work and progress towards their purpose in a transparent and open manner.

Source: Cisco.com

measurement frameworks often struggle to accommodate these divergent interpretations and may not be able to fully capture purpose (Ebrahim & Rangan, 2014). There is a need, therefore, to think carefully about how measures are introduced to capture purpose and how these are incentivised for individuals, teams, and broader units. Many aspects of purpose are qualitative, emergent, and relational and thus difficult to quantify without some loss of nuance or simplification (Micheli & Mari, 2014). Some elements of purpose can be operationalised through ESG or HR metrics, but others, such as trust, legitimacy, or (for example) purpose internalisation, are not so simple to quantify or measure (see also Chapter 8).

Further, leaders need to consider the impact of measures on behaviour. Traditional performance systems that focus on quarterly results or annual targets are unlikely to capture progress adequately when purpose tends towards longer-term thinking and sustained effort. Keeping Kerr's (1975) classic argument in his article, 'On the folly of rewarding a while hoping for B in mind,' what gets measured in organisations tends to be the thing that gets done, sometimes at the expense of what matters. There is a risk that performance metrics drive compliance over commitment, incentivising symbolic adoption rather than substantive change (Bromley & Powell, 2012). Moreover, performance measurement is not a neutral act; it is shaped by internal politics and external legitimacy pressures (Espeland & Sauder, 2007). The complexity and potential pitfalls, however, do not mean that embedding purpose into performance management systems should be avoided; fully integrating purpose into existing management and performance systems is a foundational aspect of effective purpose-led change (see also Chapter 8).

Embedding purpose in the organisation's core operations requires the use of measurable indicators beyond traditional financial metrics. While profitability remains essential, a purpose-driven organisation also needs to assess its social, environmental, and governance (ESG) contributions. A well-structured measurement system will align purpose with business success, demonstrating that social and environmental commitments enhance long-term value rather than detract from financial viability. We discuss this further in Chapter 10.

Purpose-driven organisations must consistently communicate their progress, not only to stakeholders but also within internal teams. Transparent reporting enhances trust and engagement, ensuring that employees, investors, and business partners remain committed to the purpose. A critical component of transparency is honest communication. Companies must avoid solely showcasing success. Instead, they should acknowledge challenges and outline strategies for overcoming them. Reports that only highlight achievements may be perceived as superficial, eroding trust among stakeholders (Safari & Areeb, 2020). We explore communication in more depth in Chapter 9.

Unlike financial reporting, purpose-driven reporting faces complexities related to standardisation and legitimacy (Baret & Helfrich, 2019). While financial information is regulated by well-established accounting rules, non-financial information is much less likely to be subject to an established set of regulations. Nevertheless, some broader standards do exist, such as the Global Reporting Initiative (GRI), Sustainability Accounting Standards Board (SASB) standard, and those recently developed by the International Sustainability Standards Board (ISSB). We do not suggest to blindly follow these reporting frameworks (unless mandated by the law) while losing focus on the purpose of reporting. The aim of any reported measure is to provide information that is relevant to satisfy the precise interests of all stakeholders.

Measuring and reporting on purpose-related activities should not be treated as an end in itself; rather, it becomes part of an ongoing process of continuous reflection and improvement (Schön & Argyris, 1996). Purpose-led organisations need to institutionalise learning mechanisms that enable them to adapt, recalibrate, and deepen stakeholder alignment in response to new insights. This requires a culture of adaptability in which purpose is not 'frozen' but constantly interrogated and refined through feedback and experimentation (Ebrahim & Rangan, 2014). Leaders play a central role in embedding such a culture by encouraging open dialogue, enabling critical inquiry, and supporting iterative improvements based on performance data and stakeholder engagement (Patton, 2008). When organisations approach purpose as a dynamic and evolving construct, rather than a fixed mission statement, they are more likely to sustain long-term legitimacy and responsiveness in the face of shifting societal expectations (George et al., 2023).

Concluding remarks

Purpose-driven transformation is more than just a strategic shift. It demands deep cultural change rooted in effective leadership, strong levels of engagement with the purpose at all levels, and unwavering commitment. Purpose cannot be an isolated initiative or a superficial corporate statement; it needs to integrate fully into governance, strategy, and everyday management decisions. The success of a purpose-driven transformation will depend on leaders who embody that purpose and actively drive change, ensuring that purpose is enacted authentically across the organisation.

The first step in this journey is assessing the organisation's current state, identifying gaps, stakeholder expectations, and internal barriers to purpose integration. With this foundation, leaders can define a clear road map for change. Resistance is, however, inevitable; employees, investors, and other stakeholders may push back against shifts that disrupt established ways of working. Leaders need to decide how to engage with this resistance and use it to better embed the intended purpose. Having an open dialogue and listening to concerns will help to ensure that purpose-driven strategies align with business sustainability.

Further, a culture of accountability needs to be established to sustain transformation. Leaders should implement clear performance metrics, using ESG frameworks, stakeholder feedback, and financial and non-financial indicators to assess progress. Transparency in reporting and a commitment to continuous learning will strengthen trust and ensure that purpose remains embedded in the organisation's core and evolves to meet new circumstances as they arise.

Ultimately, purpose-driven leadership is not about choosing between profit and impact; rather it is about aligning them. Leaders who challenge the status quo, model ethical decision-making, and empower employees at all levels pave the way for organisations that thrive in both business success and societal contribution. Purpose cannot be treated as an optional add-on. Instead, it should be used as a transformational force that fuels innovation, strengthens relationships, and drives long-term value.

Note

1 Authors listed alphabetically for equal contribution.

Further reading

Craig, N., & Snook, S. (2014). From purpose to impact. *Harvard Business Review*, *92*(5), 104–111. https://hbr.org/2014/05/from-purpose-to-impact

Finneran, P. (2025). Purpose driven leadership: Working together to create value for all stakeholders. *Leader to Leader*, *116*, 66–72. https://doi.org/10.1002/ltl.20878

Groysberg, B., Lee, J., Price, J., & Cheng, J. (2018). The leader's guide to corporate culture. *Harvard Business Review*, *96*(1), 44–52. https://hbr.org/2018/01/the-leaders-guide-to-corporate-culture

Rey, C., Bastons, M., & Sotok, P. (2019). *Purpose driven organizations: Management ideas for a better world.* Palgrave Macmillan. https://link.springer.com/book/10.1007/978-3-030-17674-7

References

Ackermann, F., & Eden, C. (2011). Strategic management of stakeholders: Theory and practice. *Long Range Planning*, *44*(3), 179–196. https://doi.org/10.1016/j.lrp.2010.08.001

Aguilera, R. V. (2023). Corporate purpose in comparative perspective: The role of governance. *Strategy Science*, *8*(2), 193–201. https://doi.org/10.1287/stsc.2023.0198

Almandoz, J. (2023). Inside-out and outside-in perspectives on corporate purpose. *Strategy Science*, *8*(2), 139–148. https://doi.org/10.1287/stsc.2023.0180

Andriof, J., Waddock, S., Husted, B., & Rahman, S. S. (2017). *Unfolding stakeholder thinking: Theory, responsibility and engagement.* Routledge.

Balogun, J., & Johnson, G. (2004). Organizational restructuring and middle manager sensemaking. *Academy of Management Journal*, *47*(4), 523–549. https://doi.org/10.2307/20159600

Baret, P., & Helfrich, V. (2019). The "trilemma" of non-financial reporting and its pitfalls. *Journal of Management and Governance*, *23*(2), 485–511. https://doi.org/10.1007/s10997-018-9430-z

Bromley, P., & Powell, W. W. (2012). From smoke and mirrors to walking the talk: Decoupling in the contemporary world. *Academy of Management Annals*, *6*(1), 483–530. https://doi.org/10.1080/19416520.2012.684462

Bryson, J. M. (2004). What to do when stakeholders matter: Stakeholder identification and analysis techniques. *Public Management Review*, *6*(1), 21–53. https://doi.org/10.1080/14719030410001675722

Burnes, B. (2017). *Managing change: A strategic approach to organisational dynamics.* Pearson Education.

Business Roundtable. (2024, August 16). Five years on: Corporate purpose and profit. *Business Roundtable.* https://www.businessroundtable.org/five-years-on-corporate-purpose-and-profit

Carballo, R. E. (2023). Purpose-driven transformation: A holistic organization design framework for integrating societal goals into companies. *Journal of Organization Design, 12*(4), 195–215. https://doi.org/10.1007/s41469-023-00156-8

Currie, G., & Procter, S. J. (2005). The antecedents of middle managers' strategic contribution: The case of a professional bureaucracy. *Journal of Management Studies, 42*(7), 1325–1356. https://doi.org/10.1111/j.1467-6486.2005.00546.x

Drennan, D. (1992). *Transforming company culture: Getting your company from where you are now to where you want to be*. McGraw Hill.

Ebrahim, A. (2003). Making sense of accountability: Conceptual perspectives for northern and southern nonprofits. *Nonprofit Management and Leadership, 14*(2), 191–212. https://doi.org/10.1002/nml.29

Ebrahim, A., & Rangan, V. K. (2014). What impact? A framework for measuring the scale and scope of social performance. *California Management Review, 56*(3), 118–141. https://doi.org/10.1525/cmr.2014.56.3.118

Eccles, R. G., & Klimenko, S. (2019). The investor revolution. *Harvard Business Review, 97*(3), 106–116. https://hbr.org/2019/05/the-investor-revolution

Espeland, W. N., & Sauder, M. (2007). Rankings and reactivity: How public measures recreate social worlds. *American Journal of Sociology, 113*(1), 1–40. https://www.journals.uchicago.edu/doi/epdf/10.1086/517897

Ford, J. D., Ford, L. W., & D'Amelio, A. (2008). Resistance to change: The rest of the story. *Academy of Management Review, 33*(2), 362–377. https://doi.org/10.5465/amr.2008.31193235

Freeman, R. E., Harrison, J. S., & Wicks, A. C. (2007). *Managing for stakeholders: Survival, reputation, and success*. Yale University Press.

Gartenberg, C., Prat, A., & Serafeim, G. (2019). Corporate purpose and financial performance. *Organization Science, 30*(1), 1–18. https://doi.org/10.1287/orsc.2018.1230

George, G., Haas, M. R., McGahan, A. M., Schillebeeckx, S. J., & Tracey, P. (2023). Purpose in the for-profit firm: A review and framework for management research. *Journal of Management, 49*(6), 1841–1869. https://doi.org/10.1177/01492063211006450

Gioia, D. A., Patvardhan, S. D., Hamilton, A. L., & Corley, K. G. (2013). Organizational identity formation and change. *Academy of Management Annals, 7*(1), 123–193. https://doi.org/10.1080/19416520.2013.762225

Greenwood, M. (2007). Stakeholder engagement: Beyond the myth of corporate responsibility. *Journal of Business Ethics, 74*, 315–327. https://doi.org/10.1007/s10551-007-9509-y

Groysberg, B., Lee, J., Price, J., & Cheng, J. (2018). The leader's guide to corporate culture. *Harvard Business Review, 96*(1), 44–52. https://hbr.org/2018/01/the-leaders-guide-to-corporate-culture

Heugens, P. P., Van Den Bosch, F. A., & Van Riel, C. B. (2002). Stakeholder integration: Building mutually enforcing relationships. *Business & Society, 41*(1), 36–60. https://doi.org/10.1177/0007650302041001004

Huy, Q. N. (2001). In praise of middle managers. *Harvard Business Review, 79*(8), 72–9. https://hbr.org/2001/09/in-praise-of-middle-managers

Johnson, G. (1992). Managing strategic change—strategy, culture and action. *Long Range Planning, 25*(1), 28–36. https://doi.org/10.1016/0024-6301(92)90307-N

Kerr, S. (1975). On the folly of rewarding A, while hoping for B. *Academy of Management Journal, 18*(4), 769–783. https://www.jstor.org/stable/4165235

Kotter, J. P. (1995). Why transformation efforts fail. *Harvard Business Review, 73*(2), 59–67. https://hbr.org/1995/05/leading-change-why-transformation-efforts-fail-2

Kramer, M. R., & Porter, M. (2011). *Creating shared value*. FSG.

Lyon, F., Stubbs, W., Dahlmann, F., & Edwards, M. (2024). From "business as usual" to sustainable "purpose-driven business": Challenges facing the purpose ecosystem in the United Kingdom and Australia. *Business and Society Review, 130*(512), 198–221. https://doi.org/10.1111/basr.12341

Matten, D., & Moon, J. (2008). "Implicit" and "explicit" CSR: A conceptual framework for a comparative understanding of corporate social responsibility. *Academy of Management Review*, *33*(2), 404–424. https://doi.org/10.5465/AMR.2008.31193458

May, D. R., Gilson, R. L., & Harter, L. M. (2004). The psychological conditions of meaningfulness, safety and availability and the engagement of the human spirit at work. *Journal of Occupational and Organizational Psychology*, *77*(1), 11–37. https://doi.org/10.1348/096317904322915892

McConville, T. (2006). Devolved HRM responsibilities, middle-managers and role dissonance. *Personnel Review*, *35*(6), 637–653. https://doi.org/10.1108/00483480610702700

Micheli, P., & Mari, L. (2014). The theory and practice of performance measurement. *Management Accounting Research*, *25*(2), 147–156. https://doi.org/10.1016/j.mar.2013.07.005

Noland, J., & Phillips, R. (2010). Stakeholder engagement, discourse ethics and strategic management. *International Journal of Management Reviews*, *12*(1), 39–49. https://doi.org/10.1111/j.1468-2370.2009.00279.x

Oreg, S. (2003). Resistance to change: Developing an individual differences measure. *Journal of Applied Psychology*, *88*(4), 680–693. https://doi.org/10.1037/0021-9010.88.4.680

Parmar, B. L., Keevil, A., & Wicks, A. C. (2019). People and profits: The impact of corporate objectives on employees' need satisfaction at work. *Journal of Business Ethics*, *154*(1), 13–33. https://doi.org/10.1007/s10551-017-3487-5

Patton, M. Q. (2008). *Utilization-focused evaluation* (4th ed.). Sage.

Piderit, S. K. (2000). Rethinking resistance and recognizing ambivalence: A multidimensional view of attitudes toward an organizational change. *Academy of Management Review*, *25*(4), 783–794. https://doi.org/10.2307/259206

Post, J. E., Preston, L. E., & Sachs, S. (2002). Managing the extended enterprise: The new stakeholder view. *California Management Review*, *45*(1), 6–28. https://doi.org/10.2307/4116615

Prahalad, C. K., & Ramaswamy, V. (2004). Co-creating unique value with customers. *Strategy & Leadership*, *32*(3), 4–9. https://doi.org/10.1108/10878570410699249

Quinn, R. E., & Thakor, A. V. (2018). Creating a purpose-driven organization. *Harvard Business Review*, *96*(4), 78–85. https://store.hbr.org/product/creating-a-purpose-driven-organization/R1804E?srsltid=AfmBOorraQWuhAw9XidavHXCalOO6xQ4B-xqARd-o-awEoaMl1MCszLU

Reed, M. S., Graves, A., Dandy, N., Posthumus, H., Hubacek, K., Morris, J., Prell, C., Quinn, C. H., & Stringer, L. C. (2009). Who's in and why? A typology of stakeholder analysis methods for natural resource management. *Journal of Environmental Management*, *90*(5), 1933–1949. https://doi.org/10.1016/j.jenvman.2009.01.001

Safari, M., & Areeb, A. (2020). A qualitative analysis of GRI principles for defining sustainability report quality: An Australian case from the preparers' perspective. *Accounting Forum*, *0*(0), 1–32. https://doi.org/10.1080/01559982.2020.1736759

Schein, E. H. (2010). *Organizational culture and leadership*. John Wiley & Sons.

Schön, D. A., & Argyris, C. (1996). *Organizational learning II: Theory, method and practice*. Addison-Wesley.

Senge, P., Kleiner, A., Roberts, C., Ross, R., Roth, G., Smith, B., & Guman, E. C. (1999). *The dance of change: The challenges to sustaining momentum in learning organizations*. Bantam Doubleday Bell.

Thomas, R., & Hardy, C. (2011). Reframing resistance to organizational change. *Scandinavian Journal of Management*, *27*(3), 322–331. https://doi.org/10.1016/j.scaman.2011.05.004

Todd, A. (1999). Managing radical change. *Long Range Planning*, *32*(2), 237–244. https://doi.org/10.1016/S0024-6301(99)00022-9

Torres, L. (2017). *Corporate social responsibility (CSR) and development: The case of gender equality in Latin America*. University of Nottingham. https://eprints.nottingham.ac.uk/44398/

Torres, L. (2024). Compliance. In L. Matthews, L. Bianchi, & C. Ingram (Eds.), *Concise encyclopedia of corporate social responsibility* (Vol. 54, Issue 3, pp. 18–22). Edward Elgar Publishing. https://doi.org/10.1162/002081800551280

Torres, L., Ripa, D., Jain, A., Herrero, J., & Leka, S. (2023). The potential of responsible business to promote sustainable work—an analysis of CSR/ESG instruments. *Safety Science, 164,* Article 106151. https://doi.org/10.1016/j.ssci.2023.106151

van Tuin, L., Schaufeli, W. B., Van den Broeck, A., & van Rhenen, W. (2020). A corporate purpose as an antecedent to employee motivation and work engagement. *Frontiers in Psychology, 11*. https://doi.org/10.3389/fpsyg.2020.572343

University of Cambridge Institute for Sustainability Leadership (CISL). (2023). *CISL's business transformation framework: Preliminary diagnostic*. Cambridge Institute for Sustainability Leadership. https://www.cisl.cam.ac.uk/news-and-resources/publications/cisls-business-transformation-framework-preliminary-diagnostic

Walt, V. (2021, November 21). A top CEO was ousted after making his company more environmentally conscious. Now he's speaking out. *TIME*. https://time.com/6121684/emmanuel-faber-danone-interview/

7 Purpose-led decision-making across the organisation

Catherine Tilley

Key points

- Organisations cannot claim to have embedded their purpose in their way of working until they change the way people make decisions. If decisions are focused only on economic outcomes, and do not take account of the broader aims of the purpose, the organisation will not achieve its environmental or social goals
- However, introducing the non-financial goals expressed in a purpose makes decision-making more complex. There may be tension between different elements of a purpose, and many conventional ways of decision-making are rooted either in economic theories that do not take account of non-financial considerations, or they are made using heuristics—rules of thumb in decision-making—or habits, that may not reflect the organisation's purpose
- Organisations that have made their decision-making more clearly purpose-led have done this by making changes to the formal and social structures of the organisation to influence decision-makers as well as by communicating the importance of purpose in decision-making. Individual interventions on their own cannot change decision-making: rather, organisations that succeed in embedding purpose in their decision-making take a holistic approach to the problem

Introduction

Today's decision-makers are under pressure. Customers and consumers have high expectations of businesses—especially those that have a stated purpose. They want to know that they are working with organisations that hold themselves to high standards of environmental and social care, and that pay attention to a wide range of stakeholders. At the same time, news of any missteps spreads very rapidly on social media, and it is all too easy to be accused of greenwashing or hypocrisy, leading to a loss of trust in business. We also work in a world that is changing rapidly. New technologies, environmental pressures, regulations, and a shifting geopolitical landscape mean that some decisions facing businesses are simply unprecedented and may be characterised by 'radical uncertainty' (King &

DOI: 10.4324/9781003545224-8

Kay, 2020). Tension between different stakeholders in this turbulent environment may mean that there is no one 'right' answer. Decision-makers have to navigate turbulent waters.

As we saw in Chapter 5, clear statements of purpose can provide a touchstone in these difficult times. When they are well-written, they can provide clarity about what is important in a decision (Bailey et al., 2023) and guide decision-makers. For example, the purpose statement of Dutch technology group Philips is: 'to improve people's health and wellbeing through meaningful innovation.' This is supported by a more detailed set of aims in terms of health and reach, making it relatively easy for decision-makers to focus on what really matters.

However, there is a paradox here: while a purpose statement can add clarity, introducing the non-financial aspects of a corporate purpose makes business decisions more complex. People who may previously have focused simply on a single economic outcome now need to take a more holistic view and may face trade-offs between the different aspects of the purpose and the needs of diverse stakeholders. For example, banks often have purpose statements that emphasise their role in society—HSBC's purpose statement directly references 'our customers, our people, our investors, our communities and the planet we all share' (HSBC, 2024). While this clearly highlights the importance of stakeholders, it also creates potential difficulties for decision-makers when the interests of these groups are in opposition. A bank might be interested in financing a project which produces strong economic returns for customers and shareholders, but at the expense of communities and the environment: How should the decision-makers act in this situation? Acting with purpose may thus create a problem of competing commitments for decision-makers: Who should they prioritise, and how should they manage trade-offs?

To be more than just an aspiration, an organisation's purpose needs to be enacted. People throughout the organisation need to develop new habits in their choices and behaviours. The purpose needs to infuse the entire organisation and cannot be limited to a single group or department. The purpose needs to become the guiding light for decisions taken in many functions and at many levels. As Herbert Simon wrote, 'The task of "deciding" pervades the entire administrative organisation quite as much as the task of "doing"' (Simon, 2013).

This pervasiveness creates two significant problems when organisations want to embed their purpose by changing the way people work. First, the ubiquity of decision-making in the organisation means that this new complexity cannot be 'held' in a single location to be managed by experts but has to be mastered by a wide range of people—each of whom brings different knowledge, experience, and habits of mind to their decisions. Second, although purpose statements can be helpful in signalling what is important, the existing structures for making decisions—such as the information provided, the agendas of committee meetings, or performance incentives—often direct people's attention towards economic outcomes, and may need to be revised to ensure that decision-makers are focused on the purpose.

In this chapter, we will explore the ways in which companies can start to build organisations in which their purpose informs decision-making, based on academic studies about decision-making and the experience of practitioners themselves.

The problem organisations face

For a corporate purpose to be more than just a slogan, it needs to influence the behaviour of the people in the company, shaping the way they work and the decisions that they make. While there are some organisations—particularly social enterprises—which are initiated with a specific purpose in mind, many more have adopted a purpose in recent years. Often, this is in response to a need to articulate their relationship with society and their stakeholders and to rebuild trust in the business. These companies already have established processes and routines for making decisions, which have served them well in the past. However, many conventional approaches to decision-making are based on classical economic principles which tend to prioritise economic outcomes over the social or environmental ones that underpin the purpose. As a result, it can be difficult to reconcile these methods with purpose-led decision-making.

Typically, companies adopting a purpose find that there are three aspects of their existing decision-making systems and approaches that would make it challenging to include all elements of the purpose in their decision-making. These are: the availability of data; the methods used for decision-making; and the habits of mind of the decision-makers.

1. **Availability of data**. Many of today's senior decision-makers developed their ways of working at a time when the main problem they faced was a lack of information. Today, they may be flooded with data but lack ways to turn this into insight and intelligence, and to respond in a way that is aligned with their corporate purpose. Often, corporate data systems emphasise economic data from within the company, making it hard to see what the impact of a decision might be on future stakeholders, or on stakeholders who are not well represented in the decision-making process. The data may not be easy to convert into a single unit for comparison of social, economic, and environmental outcomes. We have trouble in comparing outcomes that might be expressed in different 'currencies'—for example, it is hard to put a price on the loss of biodiversity. The Ford Pinto case (Gioia, 1992) is a classic example of this problem. Unfortunately, Ford Pinto cars tended to catch fire when involved in rear-end collisions, leading to several tragic deaths. Dennis Gioia (1992) describes how the accounting department relied on cost–benefit analysis to determine whether it was appropriate to recall all the cars. However, the cost–benefit model was not really designed to accommodate moral problems—such as the value of human life. This meant that the decision to withdraw the car from the market was delayed, and ultimately the company faced a charge of corporate manslaughter. Some of the elements that companies prioritise in their purpose statements, such as 'well-being' or 'thriving,' are too intangible and multifaceted to measure or

incorporate into information systems, leading companies to face this type of problem when they weigh up costs and benefits (see also Chapter 10).

2. **Methods used for decision-making.** Second, conventional methods for decision analysis in business tend to have biases built into them. For example, using a discounted cash flow analysis privileges the short term over the long term—which may be problematic in a company whose purpose involves a long-term time frame. Even when decision-makers take account of the needs of a variety of stakeholders, these tend to focus on 'the usual suspects'—typically, the stakeholders with the most power, urgency, or proximity, meaning that some groups—particularly those with the least access to decision-makers—will be excluded from consideration, as we saw in Chapter 4. Consequently, companies may need to change these systems if they are to embed their purpose into decision-making.
3. **Habits of mind of the decision-makers.** Finally, and perhaps most importantly, decision-makers may face cognitive challenges in adapting their decision-making to incorporate the social and environmental considerations of their purpose. Gerd Gigerenzer (2008) observed that heuristics were effective when they were well adapted to their environment, but that when the environment changed, the heuristic would need to be updated. This is the situation that many decision-makers face today: their approach has been well-developed for a corporate situation which is then radically altered by the introduction of a corporate purpose. Even with more complex decisions, for which we may not rely on rules of thumb, our habits of mind are important. Dearborn and Simon (1958) showed that we develop specific ways of looking at problems based on our professional backgrounds. Decision-makers therefore bring these habits of mind to problems that incorporate social or environmental considerations. Some organisations take advantage of this. For example, the coffee company Cafédirect includes smallholder farmers from its supply chain on the Board to ensure that they are paying attention to the issues that are relevant to this stakeholder group. The behaviour that made leaders successful before an organisation adopted a purpose may therefore need to be adapted to incorporate purpose into their decisions.

Changing these habits of mind is difficult for organisations. Decision-makers may be unconscious of their cognitive processes, or of the social factors that direct their attention. The formal structures of the organisation will also be influencing them, for example, they may have incentives that encourage a focus on shorter-term economic outcomes, and some decision-making processes—for example, standing committees—may develop biases, or 'groupthink' over time.

These problems are not independent. They arise from a system in which people have built habits of decision-making based on the type of data available, the way it is presented, the way decisions are framed, and the social pressures they face. These are all embedded in a corporate 'logic'—the 'practices, assumptions, values and beliefs that shape cognition and behaviour' (Besharov & Smith, 2014, p. 364). As companies choose to embrace a broader purpose, they need to rethink this logic and ensure that it is embedded in the formal and social systems that shape their decision-making.

What companies can do

The problems companies face are related both to formal systems, such as the information system or performance management, and social systems, such as leadership or culture, all of which can influence the way decision-makers attend to the social, environmental, and economic issues pertinent to a decision. But how do these systems interact, and how can they be changed to help people throughout the business incorporate the purpose into their choices?

There has been a wide range of research on the ways in which decision-makers manage the problem of competing commitments; for example, do we prioritise profit or purpose when both are important? Research has looked at a number of factors from the broad economic environment to the preferences of the individual and the characteristics of the decision itself. In my research, I synthesised the findings of over 100 academic publications and used these to develop the framework shown in Figure 7.1.

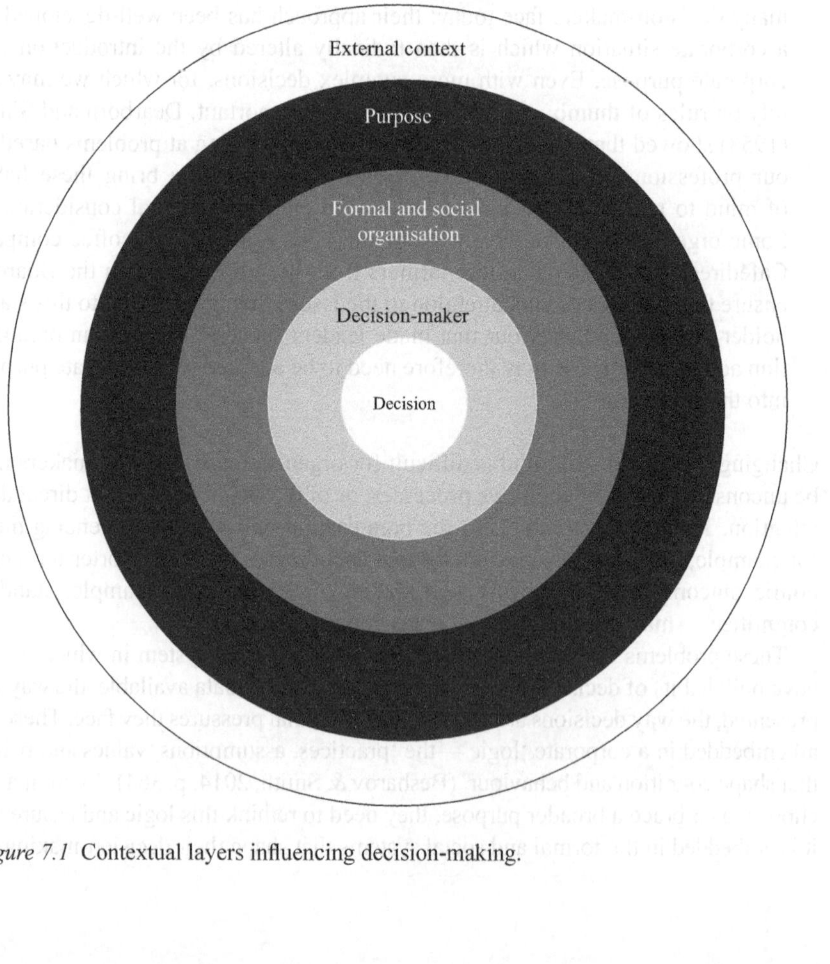

Figure 7.1 Contextual layers influencing decision-making.

This figure shows how decisions are 'nested' in a complex organisational setting. In the outside layer, we see the external context—the social, environmental, and economic pressures on the organisation. The organisation's purpose then articulates the relationship between the organisation and its external stakeholders. The purpose 'faces' outwards towards these stakeholders, but also interacts with the internal organisation, including not only its formal systems but also its social systems. These in turn influence the decision-makers by shaping and directing their attention.

If we can understand what is happening in each of these layers, we can identify the ways in which decision-makers are being influenced to think about the financial and non-financial aims of the purpose and start to identify ways in which these could be changed to embed purpose in their decision-making. To gain insight into this issue, I interviewed 40 managers at a range of large, well-established organisations about their decision-making, particularly focusing on how they manage competing commitments (such as the need to balance purpose with profit) in their work. We discussed the approaches they took to each layer of this model to shed light on the actions needed in order to connect their purpose more fully to their decision-making.

The outer layer: context and purpose

Researchers have identified a wide range of external issues, such as industry structure and stakeholders, and company-level contextual issues, such as the ownership structure or business model, which create the setting for the organisation. These are very influential in shaping corporate strategy and determining the economic behaviour of the firm. Thus, they can become embedded in the way people think about how the business works, and its ultimate aim.

Purpose is introduced in this 'outer layer' as a way of clarifying the relationship between the company and its stakeholders. If it is well articulated, it can act as a cognitive frame (Kaplan & Tripsas, 2008), influencing the way in which people pay attention to the issues in a decision. For example, the energy company Fortum states,

> Our purpose is to drive the change for a cleaner world. We are securing a fast and reliable transition to a carbon-neutral economy by providing customers and societies with clean energy and sustainable solutions.

A close reading of this statement sets some criteria for decisions. For example, the words 'fast and reliable transition' emphasise the need for energy systems to change without failing, and they also frame a potential trade-off: this is a situation where decision-makers should not compromise on reliability or speed—both are needed.

While a purpose statement can be very helpful in guiding people's attention to the important issues in their decision-making, there are two potential limitations in relying too heavily on this. First, many purpose statements are not as clear or as focused as Fortum's (Bailey et al., 2023). Second, for the purpose of influencing the way people make choices, it needs to permeate the other levels of the model, and be incorporated into the structures of the organisation, the habits of the decision-makers, and sometimes the characteristics of the decision. To do this, we need to examine each of these layers to understand how these can be influenced. We will start in the centre, looking at the decisions themselves.

The central layer: decisions themselves

Not all decisions are equal in their impact on a company's purpose. This is not to say that not all decisions matter—they do—but it is likely that there are some which more directly shape the way in which the company does purposeful work. For example, the professional services firm PwC set a purpose: 'To build trust in society and solve important problems.' The second half of this purpose depends on the type of client work that the company chooses to do, which is a choice led by the more senior members of the advisory organisation. If that group focuses more on clients who are profitable than on those who have important problems, then the company will not really be serving its purpose. Similarly, the employees choosing the technology and portfolio of power generation plants for energy company Fortum will play a critical role in determining the extent to which it can change the way it supplies energy rapidly and reliably. Companies adopting a purpose therefore need to attend to the strategically significant decisions in their businesses and consider how these might be changed by the purpose.

What these significant issues are will be specific to the company and its purpose. While the choice of client work is important to PwC because of their focus on accounting, audit, and consulting, a company in another business would need to focus on a different set of decisions to realise their purpose. For example, British responsible fashion brand Baukjen emphasises the choices they make in their supply chain, including who they work with and the raw materials that they source. These significant, shaping decisions are often visible to the broader organisation, if not all its stakeholders, and a failure to consider the purpose in them may lead people to think more cynically about the degree to which senior leaders are committed to the purpose, or the extent to which it matters in their own work.

As well as these big, purposeful decisions, smaller choices are made day-to-day throughout the organisation. Engineers and architects make decisions daily that can influence the way buildings shape our environment and our lives. Retailers can stock and display food that will make our diets healthier or less healthy. People working in customer service can help people from marginalised groups feel more included. All of these are decisions which have social or environmental impacts, and which could collectively contribute to the company's purpose.

For each of these choices, companies need to make sure that decision-makers can attend to the issues that will be important in the pursuit of purpose, and not simply economic value. The different decisions are made in different ways, but they will tend to be informed by a combination of the formal processes for decision-making—which may include the way data is provided and presented, the analytical methods used, or key processes, such as the way committees are managed—and the habits of mind of the decision-makers, which may include their experience, heuristics, and social influences. Changing the way decisions are made is complex and requires attention not only to the individuals making these choices (who need to understand the purpose and apply it to their own work) but also to the broader organisational system.

The second layer: individual decision-makers

Although organisations may have protocols for decision-making, the way these work still rests on the behaviour of the individuals. If companies are going to introduce their purpose into decisions, decision-makers need to be able to navigate the tension between profit and purpose. This ability to balance competing commitments is complex: it is not just down to the capability of the individual but can also be shaped and influenced by situational factors including available information, what the person believes they have permission to do, and the presence of stressors or situational factors that may influence the person's attention.

First, people's knowledge and understanding play a critical role. We like to believe that good business decisions are based on facts and are made rationally. However, individuals also have 'tacit knowledge'—an understanding of how to interpret a situation and how to use the data that is available effectively. Companies introducing a purpose may not have all the data they need to hand, and individuals may not have the tacit knowledge necessary to interpret what they do know. For example, the Ford Motor Company has a very aspirational purpose: 'To help build a better world, where every person is free to move and pursue their dreams.' While this is undoubtedly inspiring, it is hard to know what data would be needed to incorporate this into decision-making. Many interviewees gave examples of situations where data needed to be provided differently. One, describing a problem of tacit knowledge in the organisation, said,

> What we found before is that an awful lot of stuff didn't get thought about because the wrong people were making the decisions, well not the wrong people, but people who didn't know.
>
> (*Engineering Director—Infrastructure*)

Even when decision-makers are aware of all the issues involved in bringing the purpose into decision-making, the absence of information can be a source of stress:

> You can really feel the discomfort people have because although [our decisions] are data driven, it's not the data people have .... People are trying to figure out what's right and what's involved.
>
> (*Sustainability Director—Real Estate*)

Companies adopting a purpose can help with this by establishing frameworks or ways of structuring decisions that either allow new information to be included or enable decisions to identify the areas of uncertainty that might be problematic and discuss them. Where data is very hard to access, it can be possible to bring people with tacit knowledge or life experience to inform the decision.

A second factor that individual decision-makers face when they start to include the non-financial elements of a corporate purpose in their decision-making is their understanding of their own degree of freedom—what they are actually allowed to do—in making a choice. While job descriptions and formal structures, such as controls, may be fairly prescriptive, they do not anticipate every possible situation. This means that there is a level of managerial interpretation involved, particularly in terms of the permission individuals have to act in new situations. As one interviewee said,

> The big, tricky decisions aren't necessarily fixed by better data, systems and tools. It is more about helping people to have the confidence to make the decisions.
>
> (*Director—Public Sector*)

Because people are unclear about what they have permission to do, there can be some variation in the way decisions are made, as individuals may have different ideas about the best way to apply the purpose. People may not feel they have an explicit mandate to take a decision that involves balancing non-financial factors with economic outcomes. Decision-makers facing this type of uncertainty will tend to take their cues from the organisation, and particularly from precedents. This is one of the reasons why storytelling (which is also discussed in the organisational section and in more detail in Chapter 9) becomes an important way to help people understand what it means to work with purpose and how to include social and environmental elements in decision-making. For example, in a major UK bank, a senior leader made a short video outlining the steps they had taken to include purpose in a difficult commercial decision, and the video was made available to the staff.

Another way in which companies can encourage decision-makers to think broadly about the purpose is to be explicit about what is *not* allowed and what needs to be escalated for further discussion. One interviewee said,

> I do know what the rules are about and when to ask (for further guidance), but there's an assumption that anything outside of those boundaries is fair game . . . . The options are there to check or to act.
>
> (*Consulting Engineer*)

One company that did this reported that the escalation process helped senior leaders to understand what issues were causing difficulties for people, and where they should focus their attention.

Finally, there is a wide range of situational factors that can influence the way individuals make decisions. For example, there is an extensive body of research exploring bias. This can take two forms—the unconscious biases that arise from rapid decision-making (Kahneman, 2011) or the variation in individual preferences that people have—for example, some people may have a world view or experience that makes them more inclined to bring particular aspects of the purpose into their decision-making (James, 2000). As an example of the first type, one interviewee commented that under pressure of time, a problem could be simplified by ignoring one or more stakeholders. This might mean making a decision that did not reflect the corporate purpose (see also Chapter 4 for a more detailed consideration of different types of stakeholders). The problem can potentially be addressed through presenting decisions in a way that makes it hard to ignore the purpose—for example, by including relevant information.

A more conscious bias can arise from the individual attitudes that decision-makers hold. For example, people have a range of different beliefs about sustainable business: while some people see environmental responsibility as a cost, others may see it as a risk or even an opportunity, and these attitudes may shape the way in which they consider the environmental impacts of their decisions. This is where individuals' sense of the salience of purpose is important: when a decision-maker feels a strong sense of congruence between the corporate purpose and their own beliefs and values, it is much easier to incorporate it into their decision-making.

There are also emotional factors that can create pressures on decision-makers. One interviewee commented on the way in which a purpose can create feelings of vulnerability for leaders, as they are required to confront the difference between their current reality and the aspirations for the company. This feeling of discomfort can sometimes cause leaders to steer away from the challenges of the purpose and focus more on familiar ground. Another said,

> [Purpose] creates a case for change but also creates some fundamental cognitive dissonance. We naturally feel a desire to reduce this—especially with decisions that have been made. We find ourselves saying "if we do it that way, then everything I've been doing has been wrong."
>
> (*Strategy Director—FMCG*)

The tensions that purpose can create can be felt deeply by decision-makers and potentially become a barrier to action. This means that when a company introduces a purpose, it needs to pay attention to the factors that influence individual decision-makers: the tacit and explicit knowledge that they have; their sense of 'permission,' which drives the range of choices they will consider; and their individual thoughts and feelings. These are all influenced by the formal and social aspects of the organisation.

The third layer: formal and social organisation

In recent years, there has been a huge amount of research exploring the ways individuals make decisions, often using experiments as sources of insight. Although the work of researchers such as Daniel Kahneman (for example, his book *Thinking Fast and Slow*) has been very influential, its focus on the individual has understated the importance of the social context in shaping decision-making. Studies have shown that while individuals can and do have strong attitudes and preferences, these can be accentuated or overridden by organisational attitudes (McNamara & Bromiley, 1997). This is driven by a range of factors which are situated within the organisation itself. Some of these are formal—for example, people tend to use their job description as a guide to what they can and cannot do. But social factors—culture and leadership—also play a critical role in shaping the way we behave at work. Thus, companies that want people to use the purpose as a touchstone in their decision-making need to understand how their current formal and social structures are reinforcing (or undermining) their purpose.

Researchers have studied a wide range of factors that shape decision-making behaviour. These broadly fall into four categories: leadership, organisation design, performance management, and culture.

Leadership

There is an extensive body of research on purposeful leadership, and it is an important element of decision-making in organisations. The adoption of a purpose is almost always led from the top, even though implementation may be delegated. Leaders are important not only for the decisions they make themselves (and the issues they may face with these are discussed in the section on individuals) but also for the influence they have on the broader organisation and their ability to model the purposeful behaviour that is needed. This is a critical way to make the change visible. An interviewee at a major engineering company that was working to become more purposeful said,

> It does take behaviour at the top to make people say: "Well I can see that they care about it and so I have permission too."
>
> (*Senior Manager—Engineering Company*)

This type of visible role-modelling is clearly meaningful for people and can create legitimacy for bringing the purpose into decision-making. However, it can also create stress for the leaders. An interviewee in a senior leadership role in a bank said,

> When we are being truly purposeful, we sometimes have to be brave as leaders, and we need to role-model this. So there is a lot of leadership, and a lot of explaining previous decisions.
>
> (*Director—Retail Bank*)

This explanation process was common among the interviewees: another described it as 'thinking out loud' and a third commented on how, as a company, they had tried to create opportunities for people to see how decisions were made, so that the framing of the decision and logical approach would be clear. This created 'worked examples' which people could incorporate into their own decision-making.

Although many of the leaders interviewed were very clear about their own role in embedding corporate purpose within decisions, they also saw that purpose needed to be pervasive across the organisation. Leadership alone cannot change the entire way of decision-making. It needs to be reinforced by other corporate systems.

Organisation design

The intention behind corporate purpose is that it should apply to the entire business, not simply the top team. The design—the formal structures, systems, and processes—of an organisation is important in transferring knowledge and ensuring that priorities are communicated. There may be a designated team or task force responsible for driving the adoption or implementation of a purpose, or it may be led from a department—often HR or Communications (see also Chapters 8 and 9). However, these are not the people making all the decisions day-to-day, and so they need to ensure that the structures and processes that influence decision-making can be adapted to accommodate the corporate purpose. Organisational elements that can influence the adoption of purpose include the structures, processes, standards, and job descriptions in the organisation. Every organisation will handle these differently.

Organisation structure can be used to manage some of the tensions that decision-makers may experience between purpose and profit, either by separating elements that may be in tension, or by creating linkages between different groups. For example, one company had created a structure to ensure that pricing decisions (which needed to be fair to consumers) were made by a committee of experts, rather than by business-unit leaders who might be tempted to inflate the profitability of their units by making unfair pricing decisions. This separation meant that everyone could focus on their area of expertise without having to manage the complexity presented by a purpose that focused on the well-being of customers and communities. However, simply separating the organisation into units, each responsible for different elements of a purpose is not effective: organisations require strong linking mechanisms as well as the sharing of information. Some organisations have achieved this through, for example, cross-cutting working groups to address issues related to purpose, which can create a strong sense of inclusion. As one interviewee said, 'people now, they just all want to be on the group, and it adds a really meaningful, purposeful bit of their job. They're really excited about it' (*Senior Manager—Real Estate*).

Processes can also be vital in decision-making—for example, the Chair of a key committee in a bank made a series of changes to the way decisions were presented and discussed at the committee to incorporate the purpose more effectively, commenting that the committee: 'does two things: it provides governance, but it also changes the way we think.' In that particular case, the committee had also started to influence the way people thought by setting new standards both for preparing decisions and for the information that was brought to committee meetings. This type of revision of standards can play a critical role in shaping and directing people's attention in decision-making and can help people to see how their role contributes to the overall corporate purpose, thus focusing on what they need to do to realise this.

However, changes to these processes, standards, and job descriptions are nuanced. Not all parts of the organisation contribute to the purpose in the same way. There are likely to be some departments (and, by implication, some decisions) that play a more evident role in the purpose. For example, in the case of Fortum (the energy company focused on ensuring a 'fast and reliable' energy transition), decisions made by the engineering and portfolio teams about the type of technology to use will make a bigger difference to their progress than decisions made elsewhere in the organisation.

Several interviewees identified a particular problem with hierarchy, which is that the purpose tends to be developed and led by senior people. This means that a challenge in embedding purpose is helping everyone to make a connection between the purpose and their day-to-day work, particularly at junior levels where people may have many tasks that are not obviously connected to the purpose, and where they might not have time to stand back and consider this connection. While the NASA janitor of urban legend famously told people that he was putting a man on the moon rather than sweeping the floor, many organisations do not manage to make this connection between their purpose and the daily routine for everyone. One retail bank found that a critical element of this connection was the attention that line managers paid to purpose and so focused their efforts on ensuring that the line managers were helping frontline staff make the connections between their routine work and the higher purpose.

Performance management

One interviewee remarked, 'It is very hard for someone to change their behaviour when they are being rewarded not to do so' *(Strategy Director—FMCG)*, and performance management plays a key role in communicating expectations of purposeful behaviour to the organisation (we look some more at this in Chapter 8). There is a large body of research exploring the use of targets and incentives to signal expectations of decision-makers, and these are often used by corporations in change programmes.

One reason why HR departments are often involved in implementing a purpose is that they are generally responsible for designing and running the performance management system. This can start with articulating leadership behaviours that reflect the purpose (and values) of the company and can be used in training, development, and performance reviews. This process of articulation and development reinforces the role-modelling that is so influential in shaping decision-making, while also creating a pipeline of future leaders who are developing the skills needed to bring the corporate purpose into their ways of working.

However, implementing corporate purpose through the performance management system is not without its pitfalls. First, people often have multiple targets, and so still need to balance the potential tension between them. As one interviewee said,

> Everyone else always had an imperative that overruled it, so that the commercial guy had it as part of their role, but it was a whole lot less important than hitting their numbers.
>
> (*Strategy Director—FMCG*)

Second, the targets need to be appropriate to the operating context. One interviewee commented,

> These goals have been set by people who don't work in my area and don't work in factories and don't have much experience of social responsibility, so most of the goals were completely unachievable.
>
> (*Sourcing Director—E-commerce*)

Changing a performance management system is therefore important but can require nuance to be effective in creating an environment that directs people's attention towards the organisation's purpose.

Companies can also recognise purposeful action in a range of ways which may be separate from formal performance review—for example, nominating people for awards or setting up systems for peer-recognition. Although this type of incentive is separate from the performance management system, they maintain a focus on values and purpose. Such 'spot bonuses' seem to be a promising way of maintaining employees' interest and focus on appropriate behaviour. This can also be useful particularly in the early stages of embedding a purpose in the organisation, as it encourages peer-recognition and so builds awareness of the type of behaviour needed. The recognition system can also build case examples that are useful for communicating the value of purpose and starting to embed it into the culture, as discussed further in Chapter 9.

Culture

Organisational culture shapes the fundamental beliefs that inform our decision-making at work. It informs the social fabric of the business, and the way people interact with each other and the external environment. From a practical decision-making point of view, this can influence attitudes towards issues such as the need for consensus, sharing of information, and simply agreeing on what is important in a decision. Culture is a significant theme in the academic research on decision-making, and there have been a wide range of studies examining both fixed elements—such as attitudes, norms and beliefs, and the interventions that people make to adapt or reinforce these, such as role modelling, language, rituals, and traditions (Ardichvili et al., 2009). Culture is sticky: it has usually developed over an extended period of time and is reinforced daily through the way people work with each other, making changes to culture particularly difficult. In addition, culture is tightly woven with leadership, structure, and performance management, and is reinforced by each of these factors.

Shared beliefs are a driving force in decision-making. As one interviewee said of a purposeful decision,

> It was a relatively easy decision to make . . . there was a common sentiment that business can be a force for good.
>
> (*Marketing Director—Telecoms*)

These shared beliefs are created by stories. Many of the interviewees emphasised the importance of storytelling in creating a culture, and in many organisations, storytelling has become a vital part of embedding the purpose into the culture. Effective storytelling enables the team embedding purpose to move beyond simply communicating what the purpose is (although that's an essential start) to help people understand what it means and relate it to their jobs and decisions. As we have seen, this can give people more confidence in knowing how to apply the purpose to their decision-making. We explore this further in Chapter 9.

Stories can describe origins (for example, where the purpose came from and how) and can also perpetuate a culture of shared values. By inviting employees to share stories, it becomes clear where they see decisions at work (for example, what work to do, what clients to serve) as purposeful. Storytelling helps colleagues appreciate the actions that are being taken and can serve as a form of role-modelling. Finding the right way to communicate stories so that they are resonant may require some effort, for example, finding stories that are inclusive rather than hierarchical and telling them in an appropriate communication style. Stories need to be authentic to be effective for communicating purpose—for example, the construction company Laing O'Rourke has its origin story on its external website, which reinforces the company's purpose and explains where it comes from. The company has also collected many other stories that they use to introduce their purpose and values to their internal audiences. A major retail bank involved in this study also used stories in their internal communications in which senior leaders talked through the choices they had made and explained why they had done this. This type of 'thinking out loud' helps people to understand the way the purpose informs behaviour.

As well as *telling*, companies can also pay attention to *listening*, for example, through running focus groups which can help people to identify ways to act on the purpose (e.g. creating psychological safety so that people can do the right thing without anxiety). This type of discussion can also play a role in strengthening culture and in helping individuals feel valued. Listening processes can also help identify areas where formal structures might need strengthening to work on purpose.

A corporate purpose that is embedded effectively into a culture can enrich it and make it more distinctive as a place to work. Thus, companies that can build purpose into their culture benefit in several ways: it becomes easier to incorporate the purpose into decision-making and can become a more meaningful place to work as employees are able to experience the connection to purpose not just as communications but as a set of shared beliefs and behaviours. Because culture is hard to change, this is often reinforced through the formal systems—the structural and performance elements described earlier—to prevent the culture from sliding backwards to its pre-purpose state.

Thus, the four elements of the organisational layer are interdependent. The organisation structure, performance management system, and leadership all shape and are shaped by the culture. Embedding the purpose in these formal and social elements helps organisations to scale the influence of leaders, and to bridge what might be a divide between the people responsible for initiating and articulating the purpose, and those who work with it in their daily lives. An effective organisation shapes and directs attention towards what matters in the purpose, and helps people connect it to their work—and to their decision-making. Finally, it means that individuals working with the purpose no longer feel that they are working against the grain of the organisation by finding a balance between social, economic, and environmental outcomes.

Changing the decision-making environment

Change is difficult. However, companies that want to embed their purpose in their way of doing business need to review the way they make decisions. Figure 7.2 summarises the key influences on decision-making discussed in this chapter.

This figure shows the elements of the system that need to change to reinforce the inclusion of purpose in decision-making. As we have seen, changes at this systemic level can send a strong message to managers to incorporate the purpose in their decision-making and can shape and direct their attention. When this is done

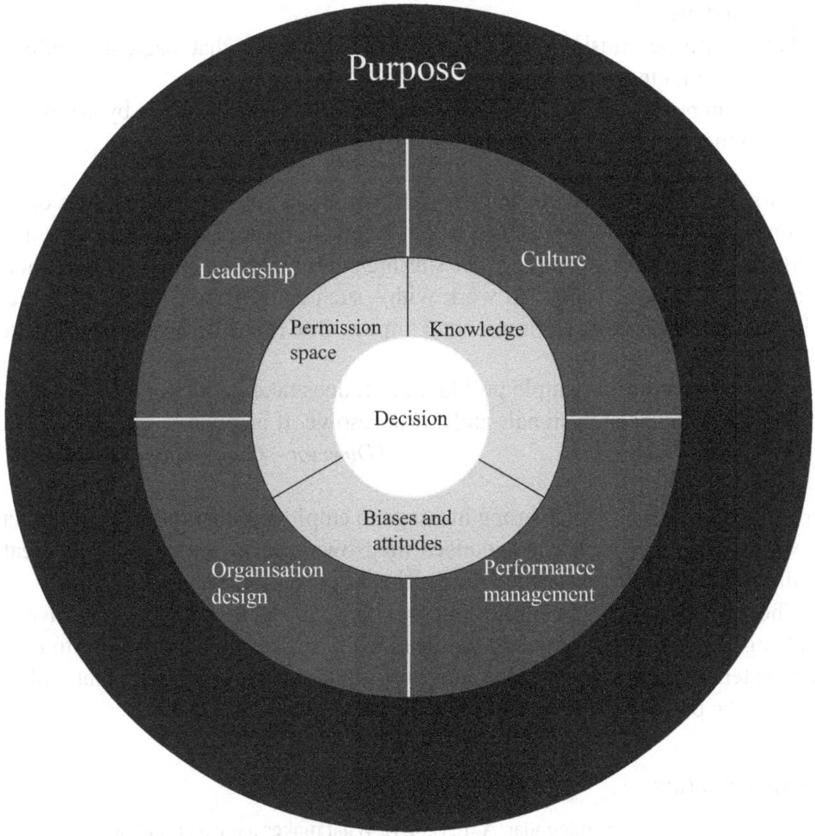

Figure 7.2 Organisational factors influencing decision-making.

effectively, it can create a feedback loop: people see purpose-informed decisions being made and feel empowered to do the same. It can also result in better decisions. Interviewees who had a clear framing of the social, economic, and environmental issues found it easier to reconcile the competing commitments between profit and purpose in their decision-making.

Architecting and driving this change is complex. While communication is vital to ensure that everyone understands the purpose and can connect it to their work, this alone is unlikely to change the full set of decisions that need to be made. Asking these five questions can help identify the actions that can help change behaviour:

1. How is the purpose currently included in the decision-making process? For example, is it represented in the information available, or the representation of stakeholders in the process? How might this need to change?
2. Do decision-makers know enough about the purpose to be able to apply it? What is encouraging them to use it, and what is getting in the way?
3. Do people feel empowered to use the purpose in their decision-making? Are there resources available to guide them if they identify trade-offs between purpose and profit?
4. Is the purpose infusing the formal and social systems that shape and influence decision-making?
5. How can purpose move beyond the high-profile decisions made by senior leaders to the daily work of everyone in the business?

The first step is to identify the high-profile, agenda-setting decisions for the organisation, and show how purpose can be embedded in these. For example, in a major consulting engineering company, the starting point was to be much more deliberate about the clients they chose to work with—such as refusing to work on projects with a negative social or environmental impact. One of the company's leaders said,

> I can't pretend it's a simple problem . . . it does take very strong leadership, and strong leadership signals and a firm resolve. It is tough.
>
> (*Director—Consulting Engineering*)

However, this stance sent a strong message to employees that they would not have to work on projects that compromised their own ethical standards. This created credibility and evidence for the introduction of purpose more broadly.

The experience of driving the purpose through decision-making at all levels challenges leaders but ultimately will lead them to develop new abilities to reconcile the tensions between purpose and profit, and to create businesses that will help to solve the problems the world faces.

Further reading

Bailey, K., Tilley, C., & Sandoghdar, A. L. (2023). What makes a great corporate purpose statement. *Harvard Business Review*. https://hbr.org/2023/09/what-makes-a-great-corporate-purpose-statement?ab=HP-hero-latest-text-1

Smith, W., & Lewis, M. (2022). *Both/and thinking: Embracing creative tensions to solve your toughest problems*. Harvard Business Press.
Todd, P. M., & Gigerenzer, G. (2012). *Ecological rationality: Intelligence in the world*. Oxford University Press. https://doi.org/10.1093/acprof:oso/9780195315448.001.0001
Thaler, R. H., & Sunstein, C. R. (2021). *Nudge: The final edition*. Yale University Press.

References

Ardichvili, A., Mitchell, J. A., & Jondle, D. (2009). Characteristics of ethical business cultures. *Journal of Business Ethics*, *85*, 445–451. https://doi.org/10.1007/s10551-008-9782-4
Bailey, K., Tilley, C., & Sandoghdar, A. L. (2023). What makes a great corporate purpose statement. *Harvard Business Review*. https://hbr.org/2023/09/what-makes-a-great-corporate-purpose-statement?ab=HP-hero-latest-text-1
Besharov, M. L., & Smith, W. K. (2014). Multiple institutional logics in organizations: Explaining their varied nature and implications. *Academy of Management Review*, *39*(3), 364–381. https://doi.org/10.5465/amr.2011.0431
Dearborn, D. C., & Simon, H. A. (1958). Selective perception: A note on the departmental identifications of executives. *Sociometry*, *21*(2), 140–144. https://doi.org/10.2307/2785898
Gigerenzer, G. (2008). Why heuristics work. *Perspectives on Psychological Science*, *3*(1), 20–29. https://doi.org/10.1111/j.1745-6916.2008.00058.x
Gioia, D. A. (1992). Pinto fires and personal ethics: A script analysis of missed opportunities. *Journal of Business Ethics*, *11*, 379–389. https://doi.org/10.1007/BF00870550
HSBC (2024, November 22). *Purpose, values and strategy*. https://www.hsbc.com/who-we-are/purpose-values-and-strategy
James, H. S. (2000). Reinforcing ethical decision making through organizational structure. *Journal of Business Ethics*, *28*, 43–58. https://doi.org/10.1023/A:100626141270
Kahneman, D. (2011). *Thinking fast and slow*. Allen Lane.
Kaplan, S., & Tripsas, M. (2008). Thinking about technology: Applying a cognitive lens to technical change. *Research Policy*, *37*(5), 790–805. https://doi.org/10.1016/j.respol.2008.02.002
King, M., & Kay, J. (2020). *Radical uncertainty: Decision-making for an unknowable future*. Hachette.
McNamara, G., & Bromiley, P. (1997). Decision making in an organizational setting: Cognitive and organizational influences on risk assessment in commercial lending. *Academy of Management Journal*, *40*(5), 1063–1088. https://doi.org/10.2307/256927
Simon, H. A. (2013). *Administrative behaviour* (4th ed.). Simon and Schuster.

8 Managing people in the purpose-led organisation

Na Fu

Key points

- Purpose-led organisations aim to integrate human, societal, and environmental goals, embedding these values into their core operations
- Achieving a shared purpose relies on managing people in a way that aligns their roles and contributions with the organisation's overarching mission
- This chapter explores how each stage of the employee life cycle, from recruitment to exit, is managed with sustainable and digital HRM practices to support purpose-driven goals
- It also addresses five key challenges in purpose-led people management, offering recommendations for purpose professionals to navigate these issues and strengthen mission alignment
- Cross-functional collaboration is critical for purpose professionals and managers, as working closely with leaders across departments ensures that the organisation's mission is consistently embedded and sustained

Introduction

Purpose-led organisations go beyond profit by embedding social, environmental, and humanistic values into their core strategy and daily operations (Bailey et al., 2024; Dhanesh, 2020; Durand & Ioannou, 2023; White et al., 2017). This approach calls for aligning people management practices with the organisation's purpose, as employees are the primary drivers of purpose through their actions, decisions, and engagements both within the workplace and externally (Bolt & Homer, 2024; Jamali et al., 2015). Sustainable Human Resource Management (HRM) offers an effective framework for fostering this alignment, given its emphasis on long-term people management strategies which prioritise employee well-being, ethical practices, and organisational sustainability (Ren et al., 2023; Wojtczuk-Turek et al., 2024). Sustainable HRM refers to

> the adoption of HRM strategies and practices that enable the achievement of financial, social and ecological goals, with an impact inside and outside

DOI: 10.4324/9781003545224-9

of the organisation and over a long-term time horizon while controlling for unintended side effects and negative feedback.

(Ehnert et al., 2016, p. 90)

It includes socially responsible HRM, green HRM (Aust et al., 2020; Diaz-Carrion et al., 2021; Wojtczuk-Turek et al., 2024), and common good HRM (Ahmad Giran et al., 2025).

- *Socially responsible HRM* refers to HR policies and practices that enhance employees' work and life quality by applying a corporate social responsibility (CSR) framework, which incorporates principles such as non-discrimination, fairness, empowerment, transparency and inclusion (Diaz-Carrion et al., 2018, 2021; Dupont et al., 2013)
- *Green HRM* focuses on HR policies aimed at addressing environmental and ecological concerns (Ren et al., 2018; Renwick et al., 2013), such as the increased pollution and reduced biodiversity in nature
- *Common good HRM* encompasses HR policies and practices that support the development of employee abilities, knowledge, skills, and attitudes necessary for addressing grand challenges in alignment with the United Nations Sustainable Development Goals (SDGs) (Aust et al., 2020; Pham et al., 2023)

As explained earlier, socially responsible, green, and common-good approaches to HRM overlap and mutually reinforce one another, collectively forming a broader, sustainability-oriented HRM framework. This chapter adopts this integrative perspective, drawing on the concept of sustainable HRM, in which sustainability principles are embedded across all HRM functions (Ehnert, 2009). Sustainable HRM supports an organisation's purpose by fostering employee behaviours that advance social, environmental, and economic objectives in alignment with the organisation's mission (Aust et al., 2020; Pham et al., 2023).

To develop and implement an organisation's purpose, sustainable HRM practices play a critical role by building employee abilities, competencies, and skills through recruitment, selection, and purpose-driven training. Additionally, they foster motivation by aligning performance management and reward systems with organisational purpose, while also creating opportunities for employees to contribute through active employee involvement. These HR functions—recruitment and selection, training and development, performance and rewards, and employee involvement—are increasingly being shaped by rapid technological advancements, particularly in artificial intelligence (AI) and people analytics. These developments offer significant benefits in efficiency, performance, and productivity, enabling HR functions to be more strategic and data-driven. Drawing on data and insights, HR analytics and digitalisation enhance decision-making that optimise workforce management across the employee life cycle from recruitment to exit (Fu et al., 2023; McCartney & Fu, 2021, 2022, 2024; Minbaeva, 2018; Rasmussen et al., 2024). In the following sections, real-world examples will be presented to illustrate how organisations have leveraged data and digital tools in each stage of employee life

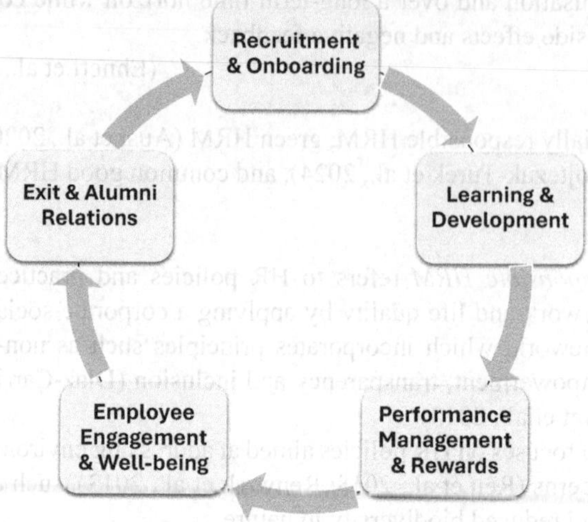

Figure 8.1 The employee life cycle.

cycle. When thoughtfully integrated, digital HRM and people analytics empower HR leaders to reinforce purpose-led goals by enhancing employee engagement, delivering real-time feedback, and ensuring HR initiatives remain aligned with the organisation's purpose.

Building on this foundation, this chapter examines how purpose-driven sustainable HRM can be integrated across the employee life cycle (see Figure 8.1), which consists of five key stages:

- recruitment and onboarding
- learning and development
- performance management and rewards
- employee involvement
- exit and alumni relations

Through a review of literature and case studies, the chapter explores how sustainable HRM practices can be effectively implemented at each stage to embed purpose-driven values and cultivate an organisational culture aligned with the organisation's purpose. Furthermore, we discuss the role of people analytics and HR technology in supporting and enhancing sustainable HRM practices, demonstrating how they contribute measurable value to the organisation's purpose. Finally, drawing the analysis of the purpose-driven sustainable HRM in the employee life cycle, five key challenges unique to purpose-led organisations are identified and discussed, including:

- balancing purpose with profitability
- integrating consistency with personalisation in people management

- leveraging technologies and analytics with a human touch
- ensuring ethical and legal use of technologies, AI, and analytics in supporting purpose-led people management
- fostering cross-functional leadership collaboration in purpose-led organisations

Practical recommendations are offered to help purpose professionals navigate these challenges, enhancing mission alignment across the workforce.

Purpose-driven sustainable HRM in the employee life cycle

Purpose-led recruitment and onboarding

Traditional recruitment primarily focuses on assessing candidates' skills and experience. However, in purpose-led organisations, recruitment begins with attracting individuals who connect not just with a job description but with the organisation's broader purpose and values (Guzzo et al., 2022; Ritchie et al., 2018). Rather than simply fulfilling a set of competencies, candidates are sought based on their natural alignment or fit with the organisation's core values and social or environmental goals. Purpose-based recruitment involves explicitly communicating the organisation's purpose in job postings, sharing purpose-driven content on social media platforms, and embedding discussions about purpose into the interview process to ensure alignment from the outset. Research shows that an organisation's social policies and programmes can enhance its attractiveness to prospective employees (Turban & Greening, 1997). This is supported by person-organisation (P-O) fit theory which refers to the extent to which people match 'in terms of their values, beliefs, and personality traits with the values, beliefs, and norms of an organization' (Saks & Ashforth, 1997, p. 396). Research shows that P-O fit raises levels of employee commitment to the organisation and improves retention rates (Kooij & Boon, 2018). Increasingly, job seekers want to understand an organisation's purpose and values to determine whether they align with their own (Hollensbe et al., 2014). This trend is especially pronounced among younger generations, who place greater importance on P-O value fit (Cennamo & Gardner, 2008; Renwick et al., 2013), making purpose-based recruitment an essential strategy for attracting and retaining talented staff.

Onboarding in purpose-led organisations similarly goes beyond the basics of task training to immerse new hires in the organisation's purpose. Programmes such as mentoring, shadowing, and structured sessions with leadership help employees understand how their roles contribute to a larger societal goal. Sustainable recruitment and onboarding emphasise transparency, fairness, inclusion, ethical practices, and long-term alignment with the organisation's purpose, allowing employees to develop an enduring connection to their workplace and find meaningfulness in their work (Bailey et al., 2024; Lysova et al., 2023). Sustainable HRM has been found to be linked with positive employee experiences and outcomes, such as believing the organisation provides a supportive and nurturing environment and improved performance (Almarzooqi et al., 2019).

Box 8.1 Patagonia

Many firms have adopted purpose-led recruitment. For example, Patagonia, the outdoor clothing and gear designer, reflects its purpose in every aspect of recruitment and onboarding. Patagonia's careers web page highlights their mission-driven approach, stating, 'We are looking for highly motivated, unconventional thinkers who love big challenges. If you are passionate about preserving wild places, value equity and belonging, and want to take action in the fight for social and environmental justice, you've come to the right place' (Patagonia, 2025). This message, focusing on the company's dedication to environmental advocacy, reinforces the alignment between potential employees' personal values and Patagonia's mission: 'We're in business to save our home planet.' During onboarding, the company prioritises alignment with environmental stewardship as much as technical skill, fostering employees' commitment to their collective goal of planetary preservation.

Digital tools enhance this recruitment and onboarding process by utilising AI to efficiently match candidates (van Esch et al., 2019; Pessach et al., 2020). For example, organisations could use AI-enabled gamification approaches to select candidates who exhibit purpose-aligned values and behaviours. Analytics in the onboarding phase can track new hires' early engagement levels, monitor cultural integration, and identify areas for additional support in aligning with the organisation's mission. These insights help HR tailor onboarding experiences that deepen employees' connections to the organisation's purpose from day one.

Purpose-aligned learning and development

Learning and Development (L&D) in purpose-led organisations is designed to align individual growth with the organisation's purpose. These programmes prioritise cultivating employees' abilities to make ethical decisions, embrace inclusivity, and contribute positively to society and the environment, rather than focus solely on technical skills or role-specific competencies. This holistic approach empowers employees to weave the organisation's values into their daily actions, fostering a work culture rooted in shared purpose and commitment.

From a sustainable HRM perspective, purpose-led L&D equips employees with the necessary tools to support the organisation's long-term mission. It adopts a proactive stance, nurturing skills and mindsets that prepare the workforce to tackle complex and evolving societal challenges. Unlike reactive training aimed solely at immediate demands, sustainable L&D emphasises adaptability, ethical foresight, and social consciousness—qualities essential for building a resilient, purpose-driven organisation. This focus strengthens the bond between employees and the organisation, encouraging sustainable practices and behaviours, which in turn lead to higher job satisfaction (Wojtczuk-Turek et al., 2024) and innovation (Fu, 2015). Employees equipped with this training are better prepared to anticipate and respond to social, environmental, and ethical challenges, ensuring alignment with the organisation's purpose.

Digital people management tools and people analytics play a crucial supporting role in purpose-aligned L&D by enabling personalised development paths and measuring the effectiveness of training initiatives (Lee & Lee, 2024). For instance, people analytics can identify skills gaps that, when addressed, reinforce the organisation's mission. If data indicates a need for enhanced expertise in environmental responsibility, the L&D team can implement targeted programmes focusing on sustainable practices. Additionally, people analytics enrich the learning experience by leveraging AI to create tailored content and delivery methods that align with individual learners' unique needs and preferences. Analytics-based dashboards can offer just-in-time feedback to learners (McIver et al., 2018), facilitating continuous improvement. Through these feedback loops, HR can adapt L&D offerings to ensure they meet both individual growth objectives and the collective purpose of the organisation.

Box 8.2 Unilever

The purpose-aligned approach to L&D is exemplified by Unilever's Future Fit Plan which focuses on equipping employees with essential skills for sustainable living, led by examples. With the purpose statement 'We are making sustainable living commonplace,' Unilever prioritises meaningful work and employee well-being, asserting that 'when we look after our people, our people look after our business.' The Future Fit Plan (Unilever, 2025) encourages employees to create personalised development frameworks that align with sustainability objectives. Through Unilever's online learning hub, degreed employees access tailored learning resources and connect with peers pursuing similar growth paths. Utilising digital technology and AI, Unilever assesses existing workforce skills and identifies future needs, ensuring employees are prepared for evolving roles in sustainability. L&D initiatives emphasise critical areas such as ethical decision-making and social responsibility, providing training on reducing environmental impact and sustainable sourcing. By embedding sustainability into their L&D practices, Unilever reinforces its commitment to purpose, empowering employees to integrate sustainable practices into their daily responsibilities and actively contribute to the organisation's purpose.

Purpose-driven performance management and rewards

Performance and rewards management in purpose-led organisations are fundamentally centred on aligning employee contributions with the organisation's purpose and values. Instead of merely evaluating employees based on traditional metrics, purpose-driven performance management emphasises outcomes that reflect ethical practices, social impact, and environmental stewardship, indicating the employees' support for their organisations' purpose. This approach encourages employees to pursue excellence in ways that contribute to the greater good, fostering a culture where purpose and performance are interconnected.

From a sustainable HRM perspective, performance appraisal motivates employees to contribute more actively to the organisation's sustainability goals (Abdelhamied et al., 2023). Holistic performance and rewards management systems not only measure and reward achievements but also cultivate an environment that supports employee well-being and purpose alignment. Research shows that performance management, along with employee voice, can lead to higher levels of employee engagement (Conway et al., 2016). In embedding purpose-driven performance management and rewards, organisations need to secure employees' involvement and voice in purpose development and implementation. For example, when designing the performance framework, employee opinions need to be sought on how to align and achieve organisational purpose in a collective way. Doing so will ensure a hybrid approach (performance and voice) in promoting and implementing organisations' purpose in their strategy. When employees' behaviours that align with organisational values are recognised, they will develop a sense of belonging and shared commitment to the organisation's purpose. By emphasising intrinsic motivators, such as personal growth, recognition, and community impact, purpose-led organisations create a workforce with shared values that is engaged and dedicated to their roles.

Digital HRM tools and people analytics are essential in optimising performance and rewards management in purpose-led organisations. Most HR systems can collect data related to performance appraisal metrics and feedback (Fernandez & Gallardo-Gallardo, 2021). People analytics can provide insights into performance trends and alignment with purpose-driven objectives, helping HR identify top performers who embody the organisation's values. Furthermore, these tools enable the personalisation of rewards, ensuring that recognition resonates with individual employees' motivations and contributions. Analytics-driven performance dashboards can facilitate ongoing feedback and performance tracking, allowing managers to recognise achievements in real time. This dynamic approach to performance and rewards management not only enhances individual accountability but also ensures that all employees are motivated to contribute meaningfully to the organisation's purpose.

Box 8.3 Danone

Danone has implemented a performance management system that ties employee objectives directly to its commitment to sustainability and social impact. In 2022, the Human Resources team conducted an intense global conversation on culture, leading to the 'Renew Danone' strategy (Danone, 2022). Employees are evaluated not only on their individual contributions but also on how their work aligns with Danone's mission: 'bringing health through food to as many people as possible' as well as Danone's HOPE values: humanism, openness, proximity, and enthusiasm. This alignment fosters a culture of accountability, where employees feel empowered to drive positive change within the organisation and beyond.

Employee involvement

Employee involvement is essential in purpose-led organisations as it connects employees with a greater cause, enhancing motivation, productivity, and job satisfaction. Employee involvement refers to the provision of opportunities for employees to speak up, to ensure managers listen to employee voices, and involve employees in decision-making (Conway et al., 2016). In purpose-led organisations, this involvement becomes even more crucial because employees are the ones who live and implement the purpose in their daily work. Their alignment with the organisation's purpose drives the success of the purpose and helps foster a motivated, satisfied, and productive workforce. By involving employees in decision-making programmes and processes, purpose-led organisations can ensure their employees feel valued and contribute more meaningfully to the organisation's overall goals.

From a sustainable HRM perspective, meaningful employee involvement extends beyond traditional productivity measures, fostering a sense of fulfilment and balance that aligns with the organisation's long-term mission (Diaz-Carrion et al., 2021). Sustainable employee involvement practices cultivate a positive work culture where employees are encouraged to contribute to social and environmental objectives, resulting in higher job satisfaction and retention (Wojtczuk-Turek et al., 2024). This can be explained by social exchange theory, which argues that employee outcomes are improved where there are mutual, trusting relationships in place between employees and their employer (Blau, 1964). The purpose-led organisation shows care for their employees, community, and environment and involves their employees in decision-making. Employees then feel a stronger sense of fit with their organisation and this, in turn, leads employees to experience a greater sense of trust and influence which helps to build healthy relationships among employees and improve overall well-being.

Digital HRM tools and people analytics again play a vital role in enhancing employee involvement in purpose-led organisations. These tools track involvement levels, identify areas for improvement, and allow HR to implement initiatives tailored to employee voice, feedback, and preferences. For example, analytics can reveal trends in employee participation in purpose-aligned training and performance management, enabling organisations to adapt strategies effectively. By using data-driven insights, organisations can develop personalised involvement strategies that resonate with employees' values and strengthen their connection to the organisation's purpose.

Managing exit and alumni relations in purpose-led organisations

Exit and alumni relations are essential aspects of people management in purpose-led organisations, shaping how employees transition out and maintain connections post-employment. Purpose-driven organisations see employee exits not merely as separations but as opportunities to reinforce their mission and values, creating lasting relationships with former employees. A growing trend is the re-hiring of 'boomerang employees' (Shipp et al., 2014), who return to the organisation after

Box 8.4 Salesforce

Salesforce exemplifies a holistic approach to employee involvement by leveraging technologies. As an AI customer relationship management platform provider, Salesforce's mission is to 'build bridges between companies and customers,' guided by five values: trust, customer success, innovation, equality for all, and sustainability (Salesforce, 2025). Salesforce develops and adopts an AI-powered platform, that is, Agentforce, to facilitate employee self-service options, enabling employees to access support and resources seamlessly across various channels. Agentforce allows the creation of AI agents capable of operating independently, retrieving real-time data, and executing tasks without human intervention. This autonomy enables employees to focus on more strategic activities while the AI agents manage routine processes. Agentforce powers AI employee service agents who assist with tasks such as onboarding, benefits management, and personalised career guidance. These agents offer a conversational interface, simplifying complex processes and providing employees with timely, relevant information. By prioritising employee involvement through digital AI tools, Salesforce fosters a culture where employees feel empowered to pursue their passions and contribute to the company's mission of social responsibility and equality.

a period away. These returning employees bring a unique combination of external experience and internal familiarity, adding significant value to the talent pool (Arnold, van Iddekinge et al., 2021; Snyder et al., 2021). Research shows that boomerang employees often outperform new hires in roles requiring extensive internal coordination, especially in environments resistant to external hires (Keller et al., 2021). Given the value of this human capital, managing employee exits effectively helps organisations maintain strong, sustainable relationships with former employees, creating a unique network of advocates and collaborators.

From a sustainable HRM perspective, exit and alumni relations strategies focus on nurturing long-term connections with former employees, recognising them as valuable brand ambassadors and potential collaborators. Alumni networks allow organisations to tap into the expertise of former employees and reinforce the shared mission that connects them. This approach fosters a sense of belonging and loyalty that extends beyond formal employment, building a community of advocates who can uphold and champion the organisation's values throughout their careers.

Digital HRM tools and people analytics are instrumental in enhancing exit and alumni relations. Machine learning and analytics, for instance, can help organisations predict employees' likelihood of leaving (McCartney & Fu, 2021). Moreover, digital tools can track alumni engagement, measure the impact of alumni networks, and identify trends that can inform future HR practices. By analysing data from alumni surveys, organisations can gauge ongoing sentiment and gather insights to better support both current and former employees. Leveraging these tools strengthens alumni relationships and ensures alignment with the organisation's purpose-driven ethos.

Box 8.5 Deloitte

A strong example is Deloitte, a global professional services firm, which manages its alumni network through the 'Colleagues for Life' initiative, guided by the principles of being 'United by a common purpose. Connected by shared experience' (Deloitte, 2025). Recognising people as their greatest asset, Deloitte has invested in the Deloitte Alumni Network, an online platform that connects over 20,000 alumni and 9,000 current staff members worldwide. Through this network, Deloitte organises events, shares thought leadership, research, and insights, and offers a diverse range of collaborative opportunities. Deloitte Global CEO Joe Ucuzoglu (Deloitte, 2025) and Deloitte Ireland's CHRO Sinead Gogan (Deloitte Ireland, 2025) emphasise that this network is 'not a network of former employees but a community of exceptional professionals who support each other's careers.' This initiative exemplifies how purpose-led organisations can sustain relationships with former employees, fostering an enduring connection to the organisation's mission and values.

Challenges and recommendations for purpose-led people management

Balancing purpose and profit via sustainable HRM

A central challenge for purpose-led organisations is balancing their mission-driven goals with profitability. While purpose-driven initiatives often result in long-term value, the upfront costs and resource commitments may appear to conflict with immediate financial performance. Leaders in sustainability and purpose-driven roles are recommended to work on integrating purpose with financial objectives, developing metrics that capture the economic, social, and environmental impact of these initiatives simultaneously. One example is the triple-bottom-line approach which includes 3Ps, that is, profit, people, and the planet (Miller, 2020). This approach not only measures traditional financial outcomes but also demonstrates how social and environmental efforts contribute to brand loyalty, customer retention, and operational efficiencies, ultimately driving profitability. We discuss this further in Chapter 10.

Purpose-led organisations can adopt HRM practices like purpose-driven budgeting and impact investing to ensure that funds are allocated to projects that align with both financial and social goals. Additionally, it is essential to communicate the business case for purpose-led initiatives clearly to stakeholders, helping them understand how these efforts create sustainable growth. For instance, promoting transparency around the cost savings from eco-friendly practices or highlighting the customer engagement gained from ethical sourcing can illustrate the financial and mission-aligned benefits. Balancing purpose and profit is not about compromise but about showing that purpose can be a strategic driver of financial resilience.

Integrating consistency and personalisation in implementing sustainable HRM

People management in purpose-led organisations requires a balance between consistency and personalisation. Consistency means that leaders manage employees in a consistent way, that is, allocating resources (Greenberg, 2009) and developmental opportunities (Greenhaus et al., 1990) across individuals in a consistent way. What is termed 'between-employee consistency' has been found to be an important factor in achieving HRM effectiveness (Bowen & Ostroff, 2004). In purpose-led organisations, communicating consistent messages across all employee groups and providing consistent support for their employees are important to ensure employees align with the organisation's purpose. While consistent policies and approaches are essential to uphold the organisation's values and mission, recognising individual differences among employees is equally important. This paradoxical mindset—embracing both consistency and flexibility—allows HR to develop practices that respect diverse motivations and personal goals while maintaining a shared commitment to the organisation's purpose (Fu et al., 2020).

To achieve this, HR can implement consistent engagement practices, such as regular consultations with employees to discuss their needs, values, and expectations. These conversations help develop a shared understanding of organisational goals and values, fostering a sense of inclusion. At the same time, during the rollout of new policies or initiatives, it is crucial to offer tailored approaches for different employee groups, adapting solutions to fit various needs and perspectives. For example, a hybrid work policy might allow flexibility for individual circumstances, such as caregiving responsibilities, while ensuring everyone has the resources to participate fully in achieving organisational goals. This balance between consistency and personalisation promotes inclusivity, helping all employees feel aligned with the organisation's mission while honouring their unique contributions and needs.

Leveraging technologies and analytics with a human touch

As mentioned earlier, purpose-led organisations use digital tools like AI and people analytics to support mission-driven goals. In doing so, balancing these technologies with a human-centred approach is crucial. Technologies enable HR to set clear metrics and measure purpose-aligned outcomes, but the human touch ensures these tools enhance, rather than replace, meaningful connections. For instance, AI can personalise onboarding and training, tailoring resources to individual needs, but pairing new hires with mentors or managers provides the personal support that builds trust. Similarly, analytics can track performance and engagement trends, yet regular one-on-one feedback sessions are essential to foster genuine employee growth.

By blending technology with personal engagement, purpose-led organisations create a work environment where employees feel valued and connected to the purpose. This balanced approach empowers people management to drive both organisational purpose and employee well-being effectively.

Ensuring ethical and legal use of technologies and analytics in people management

With digital tools and people analytics becoming core to people management, ethical and legal considerations are essential to building employee trust. Compliance with data privacy regulations, such as the General Data Protection Regulation (GDPR) and the EU Artificial Intelligence Act, is critical, requiring organisations to secure informed consent and protect employee data. Transparency around data usage is equally important, helping employees feel assured that digital tools enhance their work experience rather than merely monitoring it.

To prevent biases, algorithms should be regularly audited, and diverse data sets should be used to avoid unfair outcomes in hiring and evaluations. Human oversight is necessary to guide ethical decision-making, particularly with predictive analytics in areas like turnover risk. Insights from analytics should be applied constructively—for example, by offering support to at-risk employees instead of assigning negative labels. Embedding ethics into digital HR practices allows organisations to leverage technology responsibly, supporting engagement and alignment with purpose.

Fostering cross-functional leadership collaboration

Achieving purpose-led goals often requires dismantling organisational silos and encouraging close collaboration across departments, particularly between HR, sustainability, and other functional leaders. Purpose and sustainability initiatives are most effective when integrated across departments, creating unified efforts towards shared goals. Cross-functional collaboration ensures that purpose-driven strategies are embedded in areas such as operations, marketing, finance, and IT, enabling each function to contribute uniquely to the organisation's mission. This approach allows for sustainable supply chain practices, purpose-driven branding, and financial investments aligned with ethical values. Clear communication channels, shared objectives, and regular joint meetings help departments coordinate and measure progress on social and environmental goals, enhancing accountability. We consider purpose communication in more depth in Chapter 9.

A strategically aligned HR function plays a critical role in embedding purpose within the organisation. HR leaders should not only support but actively drive purpose-led strategies by aligning HR policies, talent management, and organisational culture with the broader purpose. This includes integrating purpose into recruitment processes, leadership development programmes, performance management systems, and employee engagement initiatives. By embedding purpose within HR strategy, organisations can attract and retain employees who are intrinsically motivated by the company's mission, fostering long-term commitment and engagement.

Leaders across functions, particularly in HR, must champion a culture of collaboration by promoting teamwork and facilitating knowledge-sharing. When leaders prioritise purpose-led goals and model collaborative behaviour, they empower teams to work cohesively and integrate purpose into daily operations. The HR function, in particular, should take an active role in ensuring that purpose is not just a statement but a lived reality, embedded in decision-making, employee experiences, and long-term strategic planning.

Conclusion

In purpose-led organisations, people management plays a crucial role in aligning employees with the purpose and values that define the organisation. From recruitment to exit, each stage of the employee life cycle can be structured to reinforce purpose, combining sustainable HRM practices with digital tools and analytics to create a supportive, purpose-driven workplace. While technologies such as AI and people analytics provide valuable insights and efficiencies, balancing these tools with a human touch ensures that employees feel genuinely valued and connected.

By fostering cross-functional collaboration, maintaining ethical use of data, and integrating both consistency and personalisation in their approach, purpose-led organisations can address the challenges inherent in balancing purpose with profitability. Ultimately, this blend of strategy, technology, and human connection empowers employees to contribute meaningfully, creating long-term value that extends beyond traditional financial metrics to drive positive social and environmental impact. In building a culture that prioritises both purpose and people, organisations can achieve sustainable success, reinforcing their mission in a way that resonates with all stakeholders.

Further reading

Aust, I., Cooke, F. L., Muller-Camen, M., & Wood, G. (2024). Achieving sustainable development goals through common-good HRM: Context, approach and practice. *German Journal of Human Resource Management, 38*(2), 93–110. https://doi.org/10.1177/239700222412408

Cappelli, P., & Nehmeh, R. (2024). HR's new role. *Harvard Business Review, 103*(5–6), 96–103. https://hbr.org/2024/05/hrs-new-role

Richards, J. (2022). Putting employees at the centre of sustainable HRM: A review, map and research agenda. *Employee Relations: The International Journal, 44*(3), 533–554. https://doi.org/10.1108/ER-01-2019-0037

Sandoghdar, A. L., & Bailey, K. (2022). *Organisational purpose What can we learn from the practitioner literature?* Research Report. King's College London. https://www.kcl.ac.uk/business/assets/research/review-of-the-practitioner-literature-on-purpose.pdf

References

Abdelhamied, H. H., Elbaz, A. M., Al-Romeedy, B. S., & Amer, T. M. (2023). Linking green human resource practices and sustainable performance: The mediating role of job satisfaction and green motivation. *Sustainability*, *15*(6), 4835. https://doi.org/10.3390/su15064835

Ahmad Giran, A. H., Muhamad Tamyez, P. F., & Fauzi, M. A. (2025). The paradoxes of sustainable HRM: Insights through the perspective of common good, green, social responsibility, and triple bottom line. *Human Systems Management*, *44*(4), 691–709. https://doi.org/10.1177/01672533251320795

Almarzooqi, A. H., Khan, M., & Khalid, K. (2019). The role of sustainable HRM in sustaining positive organizational outcomes: An interactional framework. *International Journal of Productivity and Performance Management*, *68*(7), 1272–1292. https://doi.org/10.1108/IJPPM-04-2018-0165

Arnold, J. D., van Iddekinge, C. H., Campion, M. C., Bauer, T. N., & Campion, M. A. (2021). Welcome back? Job performance and turnover of boomerang employees compared to internal and external hires. *Journal of Management*, *47*(8), 2198–2225. https://doi.org/10.1177/0149206320936335

Aust, I., Matthews, B., & Muller-Camen, M. (2020). Common good HRM: A paradigm shift in sustainable HRM? *Human Resource Management Review*, *30*(3), Article 100705. https://doi.org/10.1016/j.hrmr.2019.100705

Bailey, C., Madden, A., & Lips-Wiersma, M. (2024). Experiencing meaningful work through worthwhile contributions: A critical discourse analysis. *Human Relations*, *78*(5), 550–578. https://doi.org/10.1177/00187267241255581

Blau, P. M. (1964). Justice in social exchange. *Sociological Inquiry*, *34*, 193–206. https://doi.org/10.1111/j.1475-682X.1964.tb00583.x

Bolt, E. E. T., & Homer, S. T. (2024). Employee corporate social responsibility and well-being: The role of work, family and culture spillover. *Employee Relations*, *46*(2), 287–308. https://doi.org/10.1108/ER-02-2023-0097

Bowen, D. E., & Ostroff, C. (2004). Understanding HRM-firm performance linkages: The role of the "strength" of the HRM system. *Academy of Management Review*, *29*(2), 203–221. https://doi.org/10.2307/20159029

Cennamo, L., & Gardner, D. (2008). Generational differences in work values, outcomes and person-organisation values fit. *Journal of Managerial Psychology*, *23*(8), 891–906. https://doi.org/10.1108/02683940810904385

Conway, E., Fu, N., Monks, K., Alfes, K., & Bailey, C. (2016). Demands or resources? The relationship between HR practices, employee engagement, and emotional exhaustion within a hybrid model of employment relations. *Human Resource Management*, *55*(5), 901–917. https://doi.org/10.1002/hrm.21691

Danone. (2022). *Renew Danone: Our plan to restore growth, & drive value creation*. https://www.danone.com/stories/articles-list/renew-danone-plan.html

Deloitte. (2025). *Deloitte alumni: Colleagues for life*. https://www.deloitte.com/global/en/about/alumni.html

Deloitte Ireland. (2025). *Deloitte Alumni Colleagues for life*. https://www.deloitte.com/ie/en/about/alumni.html

Dhanesh, G. S. (2020). Who cares about organizational purpose and corporate social responsibility, and how can organizations adapt? A hypermodern perspective. *Business Horizons*, *63*(4), 585–594. https://doi.org/10.1016/j.bushor.2020.03.011

Diaz-Carrion, R., López-Fernández, M., & Romero-Fernandez, P. M. (2018). Developing a sustainable HRM system from a contextual perspective. *Corporate Social Responsibility and Environmental Management*, *25*, 1143–1153. https://doi.org/10.1002/csr.1528

Diaz-Carrion, R., López-Fernández, M., & Romero-Fernandez, P. M. (2021). Constructing an index for comparing human resources management sustainability in Europe. *Human Resource Management Journal*, *31*(1), 120–142. https://doi.org/10.1111/1748-8583.12286

Dupont, C., Ferauge, P., & Giuliano, R. (2013). The impact of corporate social responsibility on human resource management: GDF SUEZ's case. *International Business Research*, *6*(12), 145–155. https://doi.org/10.5539/ibr.v6n12p145

Durand, R., & Ioannou, I. (2023). How leaders can create a purpose-driven culture. *Harvard Business Review Digital Articles*. https://hbr.org/2023/11/how-leaders-can-create-a-purpose-driven-culture

Ehnert, I. (2009). Sustainability and human resource management: Reasoning and applications on corporate websites. *European Journal of International Management*, *3*(4), 419–438. https://doi.org/10.1504/EJIM.2009.028848

Ehnert, I., Parsa, S., Roper, I., Wagner, M., & Muller-Camen, M. (2016). Reporting on sustainability and HRM: A comparative study of sustainability reporting practices by the world's largest companies. *The International Journal of Human Resource Management*, *27*(1), 88–108. https://doi.org/10.1080/09585192.2015.1024157

van Esch, P., Black, J. S., & Ferolie, J. (2019). Marketing AI recruitment: The next phase in job application and selection. *Computers in Human Behavior*, *90*, 215–222. https://doi.org/10.1016/j.chb.2018.09.009

Fernandez, V., & Gallardo-Gallardo, E. (2021). Tackling the HR digitalization challenge: Key factors and barriers to HR analytics adoption. *Competitiveness Review. An International Business Journal*, *31*(1), 162–187. https://doi.org/10.1108/CR-12-2019-0163

Fu, N. (2015). The role of relational resources in the knowledge management capability and innovation of professional service firms. *Human Relations*, *68*(5), 731–764. https://doi.org/10.1177/0018726714543479

Fu, N., Flood, P. C., Rousseau, D. M., & Morris, T. (2020). Line managers as paradox navigators in HRM implementation: Balancing consistency and individual responsiveness. *Journal of Management*, *46*(2), 203–233. https://doi.org/10.1177/0149206318785241

Fu, N., Keegan, A., & McCartney, S. (2023). The duality of HR analysts' storytelling: Showcasing and curbing. *Human Resource Management Journal*, *333*(33), 261–286. https://doi.org/10.1111/1748-8583.12466

Greenberg, J. (2009). Promote procedural and interactional justice to enhance individual and organizational outcomes. In E. A. Locke (Ed.), *Handbook of principles of organizational behavior* (pp. 255–272). Wiley.

Greenhaus, J. H., Parasuraman, S., & Wormley, W. M. (1990). Effects of race on organizational experiences, job performance evaluations, and career outcomes. *Academy of Management Journal*, *33*, 64–86. https://doi.org/10.2307/256352

Guzzo, R. F., Abbott, J., Madera, J. M., & Dawson, M. (2022). CSR influence on job pursuit intentions: Perspectives from the lodging industry. *Journal of Hospitality and Tourism Management*, *50*, 214–222. https://doi.org/10.1016/j.jhtm.2022.02.012

Hollensbe, E., Wookey, C., Hickey, L., George, G., & Nichols, C. V. (2014). Organizations with purpose. *Academy of Management Journal*, *57*(5), 1227–1234. https://doi.org/10.5465/amj.2014.4005

Jamali, D. R., el Dirani, A. M., & Harwood, I. A. (2015). Exploring human resource management roles in corporate social responsibility: The CSR-HRM co-creation model. *Business Ethics*, *24*(2), 125–143. https://doi.org/10.1111/beer.12085

Keller, J., Kehoe, R. R., Bidwell, M., Collings, D., & Myer, A. (2021). In with the old? Examining when boomerang employees outperform new hires. *Academy of Management Journal*, *64*(6), 1654–1684. https://doi.org/10.5465/amj.2019.1340

Kooij, D. T., & Boon, C. (2018). Perceptions of HR practices, person–organisation fit, and affective commitment: The moderating role of career stage. *Human Resource Management Journal*, *28*(1), 61–75. https://doi.org/10.1111/1748-8583.12164

Lee, J. Y., & Lee, Y. (2024). Integrative literature review on people analytics and implications from the perspective of human resource development. *Human Resource Development Review*, *23*(1), 58–87. https://doi.org/10.1177/15344843231217181

Lysova, E. I., Tosti-Kharas, J., Michaelson, C., Fletcher, L., Bailey, C., & McGhee, P. (2023). Ethics and the future of meaningful work: Introduction to the special issue. *Journal of Business Ethics*, *185*(4), 713–723. https://doi.org/10.1007/s10551-023-05345-9

McCartney, S., & Fu, N. (2021). Bridging the gap: Why, how and when HR analytics can impact organizational performance. *Management Decision*, *60*(13), 25–47. https://doi.org/10.1108/MD-12-2020-1581

McCartney, S., & Fu, N. (2022). Promise versus reality: A systematic review of the ongoing debates in people analytics. *Journal of Organizational Effectiveness*, *9*(2), 281–311. https://doi.org/10.1108/JOEPP-01-2021-0013

McCartney, S., & Fu, N. (2024). Enacting people analytics: Exploring the direct and complementary effects of analytical and storytelling skills. *Human Resource Management*, *63*(2), 187–205. https://doi.org/10.1002/hrm.22194

McIver, D., Lengnick-Hall, M. L., & Lengnick-Hall, C. A. (2018). A strategic approach to workforce analytics: Integrating science and agility. *Business Horizons*, *61*(3), 397–407. https://doi.org/10.1016/j.bushor.2018.01.005

Miller, K. (2020). The triple bottom line: What it is & why it's important. *Harvard Business School Online*. https://online.hbs.edu/blog/post/what-is-the-triple-bottom-line

Minbaeva, D. B. (2018). Building credible human capital analytics for organizational competitive advantage. *Human Resource Management*, *57*(3), 701–713. https://doi.org/10.1002/hrm.21848

Patagonia. (2025). *Patagonia careers*. https://careers.patagonia.com

Pessach, D., Singer, G., Avrahami, D., Chalutz Ben-Gal, H., Shmueli, E., & Ben-Gal, I. (2020). Employees recruitment: A prescriptive analytics approach via machine learning and mathematical programming. *Decision Support Systems*, *134*, 113290. https://doi.org/10.1016/j.dss.2020.113290

Pham, N. T., Jabbour, C. J. C., Pereira, V., Usman, M., Ali, M., & Vo-Thanh, T. (2023). Common good human resource management, ethical employee behaviors, and organizational citizenship behaviors toward the individual. *Human Resource Management Journal*, *33*(4), 977–1000. https://doi.org/10.1111/1748-8583.12493

Rasmussen, T. H., Ulrich, M., & Ulrich, D. (2024). Moving people analytics from insight to impact. *Human Resource Development Review*, *23*(1), 11–29. https://doi.org/10.1177/15344843231207220

Ren, S., Cooke, F. L., Stahl, G. K., Fan, D., & Timming, A. R. (2023). Advancing the sustainability agenda through strategic human resource management: Insights and suggestions for future research. *Human Resource Management*, *62*(3), 251–265. https://doi.org/10.1002/hrm.22169

Ren, S., Tang, G., & E. Jackson, S. (2018). Green human resource management research in emergence: A review and future directions. *Asia Pacific Journal of Management*, *35*(3), 769–803. https://link.springer.com/article/10.1007/s10490-017-9532-1#citeas:~:text=DOI-,https%3A//doi.org/10.1007/s10490%2D017%2D9532%2D1,-Keywords

Renwick, D. W., Redman, T., & Maguire, S. (2013). Green human resource management: A review and research agenda. *International Journal of Management Reviews*, *15*(1), 1–14. https://doi.org/10.1111/j.1468-2370.2011.00328.x

Ritchie, G., Ashworth, L., & Bades, A. (2018). Recruiting the next generation: Applying a values-based approach to recruitment. *British Journal of Community Nursing*, *23*(5), 232–237. https://doi.org/10.12968/bjcn.2018.23.5.232

Saks, A. M., & Ashforth, B. E. (1997). A longitudinal investigation of the relationships between job information sources, applicant perceptions of fit, and work outcomes. *Personnel Psychology*, *50*(2), 395–426. https://doi.org/10.1111/j.1744-6570.1997.tb00913.x

Salesforce. (2025). https://www.salesforce.com/company/our-values/

Shipp, A. J., Furst-Holloway, S., Harris, T. B., & Rosen, B. (2014). Gone today but here tomorrow: Extending the unfolding model of turnover to consider boomerang Employees. *Personnel Psychology*, *67*(2), 421–462. https://doi.org/10.1111/peps.12039

Snyder, D. G., Stewart, V. R., & Shea, C. T. (2021). Hello again: Managing talent with boomerang employees. *Human Resource Management*, *60*(2), 295–312. https://doi.org/10.1002/hrm.22051

Turban, D. B., & Greening, D. W. (1997). Corporate social performance and organizational attractiveness to prospective employees. *Academy of Management Journal*, *40*, 658–672. https://doi.org/10.2307/257057

Unilever. (2025). *Providing skills for life*. https://www.unilever.com/sustainability/future-of-work/providing-skills-for-life/

White, A., Yakis-Douglas, B., Helanummi-Cole, H., & Ventresca, M. (2017). Purpose-led organization: "Saint Antony" reflects on the idea of organizational purpose, in principle and practice. *Journal of Management Inquiry*, *26*(1), 101–107. https://doi.org/10.1177/1056492616647481

Wojtczuk-Turek, A., Turek, D., Edgar, F., Klein, H. J., Bosak, J., Okay-Somerville, B., Fu, N., Raeder, S., Jurek, P., Lupina-Wegener, A., Dvorakova, Z., Gutiérrez-Crocco, F., Kekkonen, A., Leiva, P. I., Mynaříková, L., Sánchez-Apellániz, M., Shafique, I., Al-Romeedy, B. S., Wee, S., . . . Karamustafa-Köse, G. (2024). Sustainable human resource management and job satisfaction—unlocking the power of organizational identification: A cross-cultural perspective from 54 countries. *Corporate Social Responsibility and Environmental Management*, *31*(5), 4910–4932. https://doi.org/10.1002/csr.2815

9 Purpose-led communication

Practical steps towards defining your organisation's purpose and communicating meaningfully to your stakeholders

Joep Cornelissen and Mirjam Werner

Key points

- Communication is central to creating clarity in the minds of employees and other stakeholders about what the purpose is and in making it come alive
- Communicating the purpose begins with identifying the organisation's purpose using diagnostic questions involving a wide range of stakeholder groups and then formulating a concrete and credible purpose statement
- The initial stages involve inspirational communication aimed at activating the purpose and winning hearts and minds
- Once the purpose becomes embedded, the next phase should focus on communication that reinforces and aligns internal and external stakeholders behind this purpose
- A purpose narrative identifying the aspiration story, the character story, and the stakeholder story can be helpful in all stages of purpose-led communication

Corporate purpose and communication

Corporate purpose is oftentimes referred to as 'the [ultimate] reason for which a business is created' (Hollensbe et al., 2014, p. 3), or as 'the essence of an organization's existence' (George et al., 2023, p. 7). Although researchers in management and business generally agree on what a purpose broadly entails—namely, some societal reason or justification beyond making profits for *why* an organisation exists—the concept itself is often only defined in general terms. Within the business world, slightly different but overlapping definitions similarly exist. For example, the Davos manifesto defines a company's purpose as being about 'fulfill[ing] human and societal aspirations as part of the broader social system,' leading to 'shared and sustained value creation' with stakeholders (Schwab, 2019). The manifesto suggests that a for-profit organisation is not simply 'an economic unit generating wealth' for itself; the general idea being that, in the words of a now classic EY/Harvard Business Review study, an organisation needs to find an 'aspirational reason for being' beyond self-serving financial goals 'which inspires and provides a call to action for an organisation and its partners and stakeholders and provides benefit to local and global society' (EY/HBR, 2015). When purpose is defined in

DOI: 10.4324/9781003545224-10

this way, it becomes, in the words of Hubert Joly, the former CEO of Best Buy, 'the ultimate goal of the business, the essential reason *why* it exists, and how it contributes to the common good' (Joly, 2021b, italics in original). We examined what purpose means in greater detail in Chapter 5.

In this chapter, we define an organisation's purpose as *an aspirational, overriding reason for being which inspires and provides a call to action for an organisation that focuses on solving problems related to a specified wellbeing issue for an identified group(s) or society at large.*

This definition highlights, firstly, that purpose is a goal of sorts, but one that, as an overarching intent, is more stable and far-reaching than lower-level organisational goals. A purpose is, secondly, reflective of an organisation's own inner sense of its guiding ideals and values, and the product of its search for meaning (its 'reason for being'). Thirdly, unlike values by themselves (which may or may not be oriented towards a defined end), a purpose is always directed at an accomplishment towards which the organisation can make commitments and chart its progress. And, finally, we have highlighted how this accomplishment may be practically reachable, or not (within the lifetime of the organisation); oftentimes, its main value rather lies in the sense of direction that it provides.

In addition to the earlier definition, importantly—and in line with the focus of this book—we also primarily focus here on those organisations which state their purpose to be about contributing to a societal cause or common good that is greater than themselves. Attaching themselves to such a cause, such as reconfiguring the healthcare system (e.g. Buurtzorg), mainstreaming circular production methods (e.g. Interface), or repurposing fashion to meet climate goals (e.g. Patagonia), means that they find their purpose and reason for existence outside themselves rather than, strictly speaking, in a self-serving ethos. As such, these organisations, in terms of the purpose they pursue, tend to insert themselves into the broader industrial system or society to make headway on their *cause*, occasionally ending up as thought leaders and instigators of new movements of change in a community or industry. For these organisations, their commitments to such broader societal causes, as their declared purpose, are not only normative but also deeply existential, tying organisations and their whole being to the stewardship of such a cause. Therefore, the purposes served by these organisations are unsurprisingly often stated in generic and categorical terms around common goods or important societal challenges, rather than being specifically phrased or framed from the perspective of their own organisation or around certain core competences or capabilities (as many 'corporate' purpose statements and related marketing slogans in the business world are). Some examples include 'saving our home planet' (Patagonia), 'restoring the climate' (Interface), or 'helping sick and elderly patients live a more autonomous and meaningful life' (Buurtzorg).

As the earlier definition suggests, a sense of purpose does not stand apart from organisations, analogous to individuals and their individual purpose (Damon et al., 2003). It is rather a key component of the deep convictions that an organisation has developed about its role and purpose in society (Gulati, 2022). Purpose informs the organisation's high-level aspirations and guides the way in which it makes choices and commitments. Being or becoming purpose-led as an organisation is about bringing aspirations and commitments together, setting such organisations on a personally meaningful 'narrative quest' or 'pursuit' towards realising their self-proclaimed purpose in society. The sociologist Selznick (1957, 1992) already recognised this central role of 'walking the talk,' of organisations embodying purpose in everything that they do and communicate about (as well as to) themselves and in such a way that it leads them to stay the course. Selznick realised, following his study of the Tennessee Valley Authority, that the choices and commitments that members of an organisation make on a day-to-day basis signal to themselves and others (including stakeholders outside the organisation) their fidelity to the organisation's overarching purpose. As a result, when changes in commitments come about, whether gradual or sudden, and whether intentional or not—for instance, because some members of the organisation start to deviate from the overall purpose and make different choices—the collective pursuit of purpose becomes eroded and undermined, and might, in time, get derailed.

From this perspective, purpose is an all-embracing concept such that, in the case of purpose-led organisations, they try to embed their purpose in the organisation's culture and in the hearts and minds of all its members—so that *everyone* in effect continues to make choices and commitments in line with this purpose. Crucially, communication is central to this effort; in terms of creating clarity in the minds of employees and other stakeholders about what the organisation's purpose is (for instance, through the formulation of a compelling purpose statement) and in helping it come alive through, for example, storytelling and the sharing of examples. But, at the same time, it is also more than a set of tactical tools, channels, or content platforms through which to clearly articulate and express the purpose. In a way, purpose *is* communication all the way through: signalling through your choices

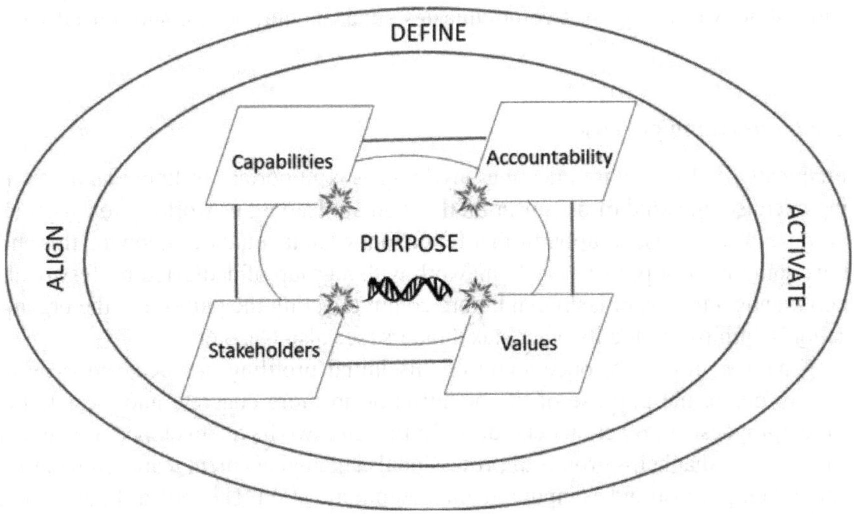

Figure 9.1 Starting with purpose: The communication model canvas©.

and commitments what you stand for as an organisation (being aware of the classic mantra that you 'cannot not communicate'), being open and transparent in all your reports, briefings and updates about your purpose alongside any business goals, and engaging in a constant dialogue with external stakeholders about shared opportunities and challenges in pursuit of the organisation's purpose. Of course, this ideal is not always met, nor easily achieved in organisations, but its importance is underscored by cases such as the one of Interface in the following text.

Being openly communicative in this way means that genuinely purpose-led organisations often see their ties to others and the communities around them as inherently more relational, cooperative, and interdependent in nature rather than conflictual, and as presenting opportunities for dialogue, joint growth, and development towards a shared 'mission.' Likewise, any conflicts, dilemmas, or challenges that arise in communicating with others are not considered as occasions for moral discipline or restraint to protect a core personal ethos (Selznick, 1992). Rather, such occasions present fault lines and contradictions that, as part of engaging with stakeholders, may be turned by the organisation into new possibilities (e.g. extensions or adjustments of its aspirations, creative 'win-win' solutions) for collaboration and communal change that may benefit all.

For example, the clothing company Patagonia realised after frequent communications with grassroots activist groups that the pursuit of its climate cause was tied to other societal issues (e.g. social justice, social equity, wealth/poverty, and colonialism), and it has been changing and evolving its commitments ever since, as well as broadening the orientation of its own purpose. The consequent expansion of the range of Patagonia's commitments (such as around creating the infrastructure for a broad range of grassroots activist groups to connect and scale with these broader climate goals in mind) reflects the company's self-aware and open-minded approach about its evolving role and purpose, as well as the very open, transparent, and public way in which it communicates about its purpose to, and indeed with, others.

Communicating purpose

Practically, before communicating anything, it is important to know as a profiling exercise, the kind of organisation that you are leading or working for, what its purpose currently is, or, upon further reflection, what it could be. Knowing this and thus doing a bit of preparatory homework with a group of dedicated professionals or a management team is crucial before communicating the purpose to the organisation's employees and its other stakeholders (see also Chapter 5).

A natural next step, once some of this initial profiling has been done, is to start to define the purpose of the organisation in more concrete and clear terms for employees and other stakeholders. To this end, we have developed a practical strategy tool that helps groups of professionals and management teams gain clarity about their purpose and as input into their communication. The tool, as highlighted, outlines the key communicative processes (define, activate, and align) involved in communicating your organisation's purpose. Box 9.1 shows the definitions of the building blocks and a set of diagnostic questions as inputs into formulating your communication strategy.

Box 9.1 Diagnostic questions for formulating a purpose communication strategy

Purpose: an aspirational, overriding reason for being which inspires and provides a call to action for an organisation that focuses on solving problems related to a specified well-being issue for an identified group(s) or society at large (see also Chapter 5).

Questions: What is the current purpose of the organisation, as captured by its purpose statement? If the organisation does not have a purpose statement, what, in your view, does the organisation see as important and as valuable to pursue and realise through all of its activities? Who does it primarily seek to serve as part of this purpose, and in what ways?

Capabilities: the skills and abilities the organisation has in developing products, in delivering services, and in partnering with others in its sector or value chain to deliver value for stakeholders.

Questions: What are the current capabilities of the organisation, and what is it particularly good at? Besides identifying current capabilities, what might the organisation with all its assets, resources, and skills be able to do in line with its purpose?

Stakeholders: the key stakeholders that the organisation has identified as important.

Questions: Who does the organisation consider its primary and secondary stakeholder groups to be? Which ones are considered most important and what value does the organisation generate for them? And are there any issues related to stakeholders (or particular under-represented or demanding groups among them) where the organisation can and wishes to make a difference in line with its purpose (see also Chapter 4)?

Values: the set of beliefs and assumptions that are shared within the organisation and its culture(s) about what is deemed important. Values may be both lived and desired and may involve shared beliefs ('respecting diversity') as well as views on what are deemed to be valuable ways of working ('acting with integrity' and 'commitment to innovation'), based on assumptions about what the organisation is purpose-wise reaching for and what will drive that.

Questions: What are the values of the organisation, as captured in an organisation's official value statements, and as practiced and lived in the organisation as well? Which of these values seems to be overriding principles by which people in the organisation seem to act? How are these linked to the purpose?

Accountability: the target metrics and incentives that the organisation uses to measure what it wishes to achieve; how the organisation evaluates progress, and how it has created and distributed incentives to make the organisation's purpose tangible (see also Chapter 10)

Questions: What metrics and incentives are currently in place within the organisation? How are people incentivised to make decisions in line with the company's purpose, if at all? What do the incentive structures themselves convey about what is deemed important within the organisation? (see also Chapter 8)

As a first step, the aim is to formulate a purpose for the organisation (see Box 9.1) and to diagnose to what extent the pursuit of this purpose is already reflected in and supported by each of the building blocks. As a group of professionals or management team, the objective is to lay this out as clearly as possible and identify possible leads for internal development and communication with employees, and for external communication purposes with other stakeholders as well. We examined purpose development in further depth in Chapter 5 in the context of the recent national and international standards that have been proposed; here, we focus on purpose development in the context of communication.

It may be useful to highlight this initial 'defining' step with a well-known case. In 1994, Interface, a carpet manufacturer, began a journey to become a sustainability-led, circular company. Ray Anderson, the company's founder and CEO, had read the book *The Ecology of Commerce* in that year, which, as he later described, 'was a spear through the chest' experience. He realised that rather than plunder the Earth with his petrol-intensive carpet business, the company needed to make a U-turn and embrace an *environmental purpose* as its cause. Struck by the revelation that the company's business model was unsustainable and would leave the planet worse off for future generations (whom he now considered as *primary but previously neglected stakeholders* besides its employees, suppliers, customers, and investors), Anderson encouraged his company to use its *capabilities* in design and engineering to begin searching for innovative, sustainable solutions.

He boldly formulated a new purpose for the organisation:

> To be the first company that, by its deeds, shows the entire industrial world what sustainability is in all its dimensions: People, process, product, place and profits—by 2020—and in doing so we will become restorative through the power of influence.

The associated targets and metrics (*accountability*) that he and his management team set for the company were similarly bold and were meant to 'eliminate any negative impact Interface has on the environment by 2020': zero waste, zero greenhouse gas emissions and net water use; 100% renewable energy; and a closed technical loop, via product recycling and take back and by using 100% recycled or bio-based materials for its products (Cornelissen, 2023).

Anderson and his team also specified other sustainability goals related to transportation, stakeholder well-being, and new business models that he thought would 'redesign commerce.' For example, Interface designed and mainstreamed a recycling process, whereby its carpet tiles were completely recycled, as an example of a circular business model. The company introduced and perfected processes such as these to make progress towards its ambitious targets (*accountability*), and in doing so also signalled to others in industry the possibility of a circular economy with zero waste and negative environmental impacts of business processes. Finally, to support the pursuit of its new purpose, Anderson and his team realised early on that the company needed to foster a culture characterised by open innovation and by a set of *values* that conveyed a tolerance for experimentation and failure. The company in turn set up an intensive programme of training and internal communication to socialise employees into the new purpose and to encourage specific values around experimentation, acceptance of failure, mutual respect, and personal development and growth.

Box 9.2 Purpose statements—Best practice guidelines

Many organisations have captured their purpose in a short statement (often one sentence long, ranging between 5 and 12 words) that starts with an action-based word that declares their commitment towards an aspiration (see Box 9.1 and also Chapter 5 for examples). Like other kinds of slogans or statements, a purpose statement should not only clearly articulate the organisation's overall aspiration but should also uniquely identify and define the organisation such that the resulting statement is considered by its employees and other stakeholders as authentic and credible.

When you are involved in drafting or redrafting a purpose statement, it might be useful to first bear the following strategic considerations in mind:

- The power of the **positive question**: purposeful change is most likely successful when it elevates and extends strengths and shared beliefs, and broadens and builds on an organisation's capacity to make a difference
- Find (or renew) your purpose through fostering **dialogue**: finding common ground by sharing discoveries and possibilities, dialoguing and debating, and finally getting to the point where everyone can say, 'Yes this is something that we all value and should aspire to. Let's make it happen'
- The **magic of words**: purpose statements will more likely inspire and galvanise when they are worded in such a way that they connect the head and the heart, emotionally moving members of the organisation and stakeholders towards a joint purpose

Then, when you get down to the business of drafting the purpose statement, incorporate the following best practice guidelines:

- Make it a short statement (up to 5–7 words only, or one short sentence)
- Start with an action verb
- Refer to an outcome for an addressee (stakeholder, beneficiaries, world)
- The statement may or may not refer back to a company's products/services, capabilities, or the industry/field that it is in
- Employ rhetoric (figurative language, idioms, and iconic words) BUT wisely
- Connect the head and the heart
- Express the overarching 'reason why' that is delivered in what you do
- Ensure the statement is tangible, yet captures a universal and broader (social) good
- And like any slogan, it should be evocative and memorable

When, after a few rounds of drafting, you have a purpose statement, you may want to check it with others in the organisation and with the organisation's external stakeholders. Any feedback that you get subsequently can be helpful input for further revision.

Concreteness check: after attempting to craft your purpose statement, ask yourself and your stakeholders the question: Is our Purpose Statement *concrete and tangible* enough?

Authenticity check: after attempting to craft your purpose statement, ask yourself and your stakeholders the question: Is our Purpose Statement credible and can it be *instantly felt,* apart from being understood?

A direct outcome of the above-mentioned diagnostic questions is the definition of a compelling purpose statement (or revising an existing one based on your analysis) as well as an identification of specific development and communication steps that will further align each of these building blocks to the declared purpose. The overall aim with identifying such tactical development and communication steps is to 'activate' the organisation's purpose by further infusing the different building blocks with the purpose as an animating force. Leadership and communication are central to this 'activation' step to make the purpose come alive and inspire employees across the organisation as well as key stakeholders to align themselves with the shared aspiration. This is important as well to inform decision-making, as we saw in Chapter 7.

Returning to the case example of Interface, core to activating the purpose in those early days was Ray Anderson himself, who tirelessly communicated the new direction for the company to all staff. He gave more than 1,000 speeches at the start of the change process, to people inside and outside of the organisation. He told his staff of the importance of collectively scaling 'Mount Sustainability,' a metaphor that cleverly captured the challenges and milestones ahead and, as part of more extended narrative (see Box 9.3), articulated the journey ahead.

Anderson was a gifted speaker; with his Texan accent and direct way of communicating, he inspired people and convinced them of the radical change that he was proposing for his company and for the industry at large. A good example of his persuasive style is the TED Talk ('The business logic of sustainability') which he gave in 2009 (Anderson, 2009). In the video, he skilfully alternates between different ways of framing the importance of sustainability for business; as marking social progress, as making real business sense (because of cost savings and new market opportunities), and as morally the right thing to do for your fellow human beings and for future generations (as primary stakeholders) (Anderson, 2009). Besides extensive communication, Anderson and his team also realised that some adjustments around each of the building blocks had to be made. For example, in terms of *capabilities*, they felt that to become a sustainability thought-leader in industry, the company had to work more openly and collaboratively with others across its value chain and with sustainability experts. Amidst other initiatives that they took, Interface created an 'Eco Dream Team' made up of authors, activists, scientists, and entrepreneurs to bring together knowledge from different disciplines and initiate a new way of collaborating and partnering with others to find novel ideas for circular business models as well as share the resulting knowledge as broadly as possible (as opposed to keeping it within the boundaries of the organisation as commercially proprietary knowledge). Relatedly, instead of seeing competitors in antagonistic terms, Interface realised that they had to include them as primary *stakeholders,* with whom they could collaborate (rather than compete) to create 'win-win' solutions and the possibility of more systemic change that may benefit all.

When, as a final step, as an organisation you reach the situation where each touchpoint, decision, and experience of stakeholders starts to faithfully reflect the organisation's purpose, the job shifts from inspirational communication aimed at 'activating' a purpose and 'winning hearts and minds' to more broad-based communication programmes and content platforms that build relationships with key stakeholders around the purpose with the aim of ensuring that they keep 'aligning' themselves behind the purpose. When everything is starting to become more aligned in this sense, the organisation can build confidence and trust internally and externally to sustain behaviour in line with its purpose. The focus accordingly shifts to the linkages between the building blocks as part of the system as a whole; and ensuring that the organisation keeps 'walking the talk' (backs up its rhetoric with substantive commitments) and 'talking the walk' (communicates about its purpose-driven commitments and achievements in an ongoing and meaningful manner with its stakeholders).

Going back to our example of Interface, the company in those early years managed relatively quickly to align the building blocks with its new purpose, setting it on a path of becoming a circular company and sustainability thought-leader. The purpose became so broadly shared and lived by employees throughout the company that it became naturally sustained over time. In fact, because of this alignment, Interface made such significant progress on its purpose that, despite the audacity of its initial aspiration, it reached its goals ahead of time. By 2020, the company had halved its energy usage, powered its plants with 87% renewable energy, cut water usage by 90%, sourced recycled biobased resources for virtually all its products, and reduced greenhouse gas emissions by 95%. As a result, the current management team of the company decided to revise the company's purpose and state a new aspiration beyond 2020, titled 'Climate Take-Back.' As the title suggests, Interface is envisioning a new strategic direction for the company (still in line with its core purpose of being a restorative influence), where rather than minimising its negative impact on the environment and bringing it close to net zero, it would instead be a builder and restorer of natural ecosystems. The idea is to create solutions that would help combat climate change and take carbon out of the atmosphere. One example of this new iteration of its purpose is to design 'factories as forests'—to use its capabilities to build facilities that, like the ecosystem that it replaces, contribute to the natural environment, in terms of fostering a wildlife habitat and supporting water storage and purification, temperature cooling, and other biological processes. With its revised purpose, Interface hopes to chart a new and unprecedented trajectory for an industrial company and, as with its previous iteration, in doing so, it hopes to inspire its competitors and others in the industry so that sustainability becomes truly mainstream.

One particularly helpful communication tool that Interface used is a purpose narrative (see Box 9.3). An official, written-out narrative is helpful not only for activating a purpose but perhaps even more importantly for aligning employees and stakeholders, because it gives them a continuing sense of how all their actions are part of a sustained and authentic pursuit of a common purpose. Narratives also convey the organisation's purpose and values in action, helping employees and other stakeholders *feel* what matters as well, rather than just thinking about it or telling others what matters. Because narratives allow a purpose to be expressed not as abstract principles but as lived experience, they have the power to move others and enlist them in the common pursuit.

Box 9.3 Activating the organisation's purpose through a purpose narrative

A purpose narrative is a powerful way of activating the purpose inside and outside the organisation. Like most narratives, it involves a storyline around a core character (the organisation) who, in the face of challenges, is called upon to make the right choices leading to some desired outcome (as the plot).

To craft a compelling purpose narrative, we believe it helps to write it from the perspective of the organisation as the focal actor and think of ways to thematise and plot a narrative from that perspective: How did the current purpose come about? Was it already part of the company's history and culture? How does it reflect the organisation's character? What outcome or progress would the company like to see?

Then, to give it more concrete shape, it is helpful to break the overall narrative down into three story parts and to work those out in more detail. You may want to write up each story part more freely at first, but then when the different parts take shape, bring them together and work towards a composite narrative of about 7–12 lines.

Aspiration story

What is the change you want to make in the world?

This part of your narrative communicates the urgent challenge the organisation is called upon to face now, and the aspiration it has to make a difference. It focuses on a cause, issue, or general problem that requires action, the hope for that action, and the choices the organisation is calling upon others to make as well.

An example of an Aspiration Story, taken from Rabobank's overall purpose narrative, is the following paragraph:

> Each generation faces its own unique challenges—big and small. And in every era, people have joined forces in search of solutions to these diverse issues. Making our lives and businesses more sustainable is

> the number one challenge of our time. The world is facing three major global transitions: food, climate, and energy; as well as the challenge of transitioning to a more inclusive society.

Character story

Why are you as an organisation called to make that change and have this aspiration; what specific experiences have shaped your Character Story?

This part of your narrative focuses on choices and commitments, on moments in the organisation's history when its ideals and values were clearly conveyed. The power in your Character Story is to reveal something of the organisation and its values—including the key shaping moments in its (recent) history and how and why these link to its current aspirations.

An example of a Character Story, taken again from the Rabobank purpose narrative, is the following:

> Rabobank has been doing this [joining forces through cooperation] for almost 125 years. It started with inventive farmers: what they could not do on their own, they were able to achieve together. This farmers' cooperative turned into a bank, which today is embedded in society thanks to its members. We have worked in the same way since our foundation: together, as a cooperative. And we will continue to do so, because the future, too, is cooperative. That is why we help strengthen communities and make the local living environment across the Netherlands greener.

Stakeholder story

What story can you tell that will help others understand why you aspire to make that contribution and foster change: What is your Stakeholder Story?

A Stakeholder Story communicates as part of your narrative why the broader stakeholder community should care about your aspirations and your quest as an organisation. The key is to focus on telling stories about specific people and specific moments of choice or action that invite others to join in and support you in your quest. Depending on the stakeholder audience, you can then also adjust your overall narrative with more specific stakeholder-specific appeals and storylines.

An example of a stakeholder story from the Rabobank purpose narrative is the following:

> We embrace innovation, growth and sustainability, and invest in joint solutions that benefit everyone: our clients, their neighborhoods and the world. We also invest part of our profits in society—sometimes in your neighborhood, sometimes further afield, for example by helping

African farmers. In this way we support and learn from each other. Together, we make sure that we stay financially healthy and that our villages and cities remain attractive and dynamic. We are involved in what is going on in society. We contribute to strengthening our shared living environment and making it future-proof—so that people can make the choices that matter most to them, both now and in the future.

Finally, once all the parts are written up, and a coherent purpose narrative emerges, you can think of adding a temporal element that shows, over time and through resolve and a strengthening of commitments, the organisation has become (or is becoming) more infused by its values and more purpose-led. And finally, it may be useful to think of adding specific story elements that create resonance and give fidelity to the story (e.g. vivid or concrete descriptions), further fostering its believability and potential impact.

Once you have written out a purpose narrative for your organisation (using the steps mentioned later in the text), it can be a powerful component of your purpose communication that can feature on your website but can also be repurposed and repackaged for organisation-wide communications (mailings, newsletters, briefings, job adverts, etc.) as well as for stakeholder specific communications (such as briefings for journalists, calls with financial analysts, prospectus for investors, and brand packaging).

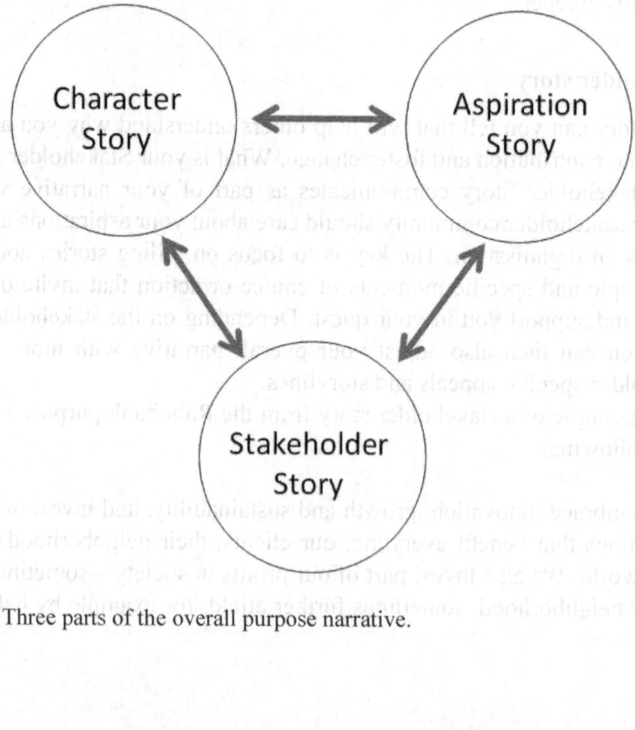

Figure 9.2 Three parts of the overall purpose narrative.

The power of communication

In this chapter, we have offered a working definition of the concept of purpose and have outlined what it takes for organisations to become truly purpose-led. Using examples from existing cases, we have highlighted how a sense of purpose is first and foremost linked to an organisation's sense of the cause that it serves in society, and what aspirations its members collectively deem as important and meaningful to pursue. We have also outlined a tool to help those working for organisations to chart the development of their organisation's purpose (or the revision of their current one). Through this tool, they can use the power of communication to make the purpose come alive amongst employees within the organisation and become recognised and supported by external stakeholders as well.

We specifically featured the role of purpose statements and purpose narratives as powerful means to animate everyone inside and outside the organisation to work together in pursuit of a purpose. But beyond such statements and narratives, the whole process of forming a purpose for an organisation is, of course, a communicative one at heart; of constantly dialoguing together, of articulating to one another who and what is meaningful to serve, and of identifying together how we can work towards a joint aspiration. In this sense, constant consultation and dialogue through open town halls, brainstorms, team meetings, hackathons, and bottom-up sessions of listening to employees (as evidenced by cases such as Interface) are absolutely key to ensure that employees feel involved and personally embrace the purpose as meaningful (and effectively become stewards of the purpose themselves), and so that opportunities and possibilities for becoming more purpose-led as an organisation are mutually defined (Cornelissen, 2023).

A similarly open, interactive process of engaging with external stakeholders likewise ensures that a purpose is more likely to become shared and recognised by them, where they might even join the organisation in its quest. The focus with such stakeholder engagement is thus not merely on shaping opinions or perceptions, but on the organisation being transparent about its purpose and inviting external stakeholders in, so that they feel involved, consider the organisation's actions as 'authentic,' and potentially become genuine advocates of the organisation's pursuits. If purpose is simply announced top-down by leaders or senior managers as a slogan or, worse, hidden somewhere on the corporate website without involving employees or external stakeholders, it is far less likely to bear fruit. This is particularly true when none of the building blocks that we have mentioned are altered to align with the newly declared purpose. An organisation has to demonstrate commitment in pursuit of its purpose, effectively putting its money where its mouth is, and being authentic and transparent about this in all internal and external communications.

Further reading

AW Page Society. (2012). *Building belief: A new model for activating corporate character and building belief.* https://knowledge.page.org/wp-content/uploads/2019/02/Full_Report.pdf

Cornelissen, J. (2023). *Corporate communication: A guide to theory and practice* (7th ed.). Sage Publications.

Joly, H. (2021a). *The heart of business—Leadership principles for the next era of capitalism*. HBS Press.

Mayer, C. (2018). *Prosperity: Better business makes the greater good*. Oxford University Press.

References

Anderson, R. (2009). *The business logic of sustainability* [Video]. TED Conferences. https://www.ted.com/talks/ray_anderson_the_business_logic_of_sustainability?language=en.

Cornelissen, J. (2023). *Corporate communication: A guide to theory and practice* (7th ed.). Sage Publications.

Damon, W., Menon, J., & Bronk, K. C. (2003). The development of purpose during adolescence. *Applied Developmental Science*, *7*(3), 119–128. https://doi.org/10.1207/S1532480XADS0703_2

EY/HBR. (2015). *The business case for purpose*. HBS Publishing. https://assets.ey.com/content/dam/ey-sites/ey-com/en_gl/topics/digital/ey-the-business-case-for-purpose.pdf

George, G., Haas, M. R., McGahan, A. M., Schillebeeckx, S. J. D., & Tracey, P. (2023). Purpose in the for-profit firm: A review and framework for management research. *Journal of Management*, *49*(6), 1841–1869. https://doi.org/10.1177/01492063211006450

Gulati, R. (2022). *Deep purpose: The heart and soul of high-performance companies*. Penguin.

Hollensbe, E., Wookey, C., Hickey, L., George, G., & Nichols, C. V. (2014). Organizations with purpose. *Academy of Management Journal*, *57*(5), 1227–1234. https://doi.org/10.5465/amj.2014.4005

Joly, H. (2021b, October 28). Creating a meaningful corporate purpose. *Harvard Business Review*. https://hbr.org/2021/10/creating-a-meaningfulcorporate-purpose.

Schwab, K. (2019). *Davos Manifesto 2020: The universal purpose of a company in the fourth industrial revolution*. https://www.weforum.org/agenda/2019/12/davos-manifesto-2020-the-universal-purpose-of-a-company-in-the-fourth-industrial-revolution/

Selznick, P. (1957). *Leadership in administration*. Harper and Row.

Selznick, P. (1992). *The moral commonwealth: Social theory and the promise of community*. University of California Press.

10 Measuring and reporting on purpose at a time of crisis and unravelling consensus

David Aikman and Marc Lepere

Key points

- Unravelling climate consensus: The global consensus on decarbonisation is breaking down, with increasing political polarisation. While some advocate for urgent climate action and greater emphasis on corporate purpose, others challenge its necessity and cost, trusting future technologies to provide solutions
- Limitations of climate finance and disclosure: The evidence on the effectiveness of financial markets, corporate disclosures, and voluntary commitments such as corporate purpose in driving decarbonisation is mixed. Despite expanded disclosure frameworks and growth in sustainable finance, concerns about greenwashing, mispricing of risks, and regulatory inconsistencies persist
- Intensifying financial risks of climate change: The financial impacts of climate change, including rising insurance costs, increasing litigation, and economic losses from extreme weather events, are intensifying. Businesses face growing financial and operational challenges due to these risks
- Strategic decision-making for business leaders: For corporate leaders, the challenge of climate risks is not whether to act, but how to act strategically and with purpose. We offer eight decision-critical questions to help leaders integrate climate resilience into operations, risk management, and financial planning in an increasingly uncertain climate landscape

Introduction

The fragile consensus on the need to decarbonise the global economy is unravelling. Climate policy has become a fault line in an increasingly polarised political landscape, with two opposing camps emerging. On one side, a substantial coalition continues to sound the alarm on the climate emergency, urging lawmakers and businesses to accelerate the transition to net zero. On the other, a growing and vocal movement frames climate action as an unnecessary, costly distraction—arguing that climate risks are exaggerated and that future technologies, not current policies, will provide the solutions. This breakdown in consensus is no longer just a theoretical debate; it is reshaping the business and financial landscape in real time, making long-term strategy increasingly fraught for corporate leaders.

DOI: 10.4324/9781003545224-11

Alongside this political shift, several realities have become inescapable. First, the mechanisms designed to deliver a 'just transition' to net zero are faltering. The United States has once again distanced itself from the Paris Agreement; the Glasgow Financial Alliance for Net Zero (GFANZ)—a global coalition launched during the COP26 of financial institutions committed to aligning their portfolios with net-zero emissions by 2050—is collapsing as financial institutions retreat from climate commitments; greenwashing[1] has given way to 'greenhushing' as companies seek to avoid scrutiny; and financial markets still struggle to price climate risk effectively. Yet, against this backdrop of policy uncertainty and inertia, the physical impacts of climate change are accelerating, and global emissions continue to rise with no imminent sign of reversal. In just the past six months, catastrophic floods have devastated parts of Valencia, and wildfires have ravaged Los Angeles. Scientists have warned that the hopes of keeping global warming below 1.5°C above pre-industrial levels have been all but been extinguished after 2024 was the first calendar year to see average temperatures breach that critical threshold. We explored these climate changes in more depth in Chapters 1–3.

This chapter examines the financial, regulatory, and strategic uncertainties facing corporate decision-makers as they navigate climate-related risks. Firms are responding in different ways: some are taking a compliance-driven approach focused on meeting regulatory expectations; others are integrating climate risk into core strategic decision-making. A growing subset is turning to corporate purpose as a lens for action—framing environmental and social commitments not merely as obligations but as opportunities to build stakeholder trust, attract capital, and enhance long-term value.

However, the extent to which corporate purpose alone can drive widespread, systemic change remains a subject of debate. A key theme of this chapter is that durable progress will require climate action to be underpinned by commercial incentives. The role of corporate purpose lies at the heart of a long-standing debate in finance and management. The shareholder primacy view, which prioritises maximising shareholder returns, often clashes with stakeholder theory, which argues that corporations should consider the interests of all stakeholders, including employees, customers, and the environment. This chapter will examine how these competing perspectives shape the understanding and implementation of corporate purpose in the context of climate change. While many companies are embracing corporate purpose, there is a risk of 'purpose-washing,' where purpose statements are used for the sake of public relations without genuine commitment to change. This can erode trust and undermine the credibility of the corporate purpose movement.

We begin by outlining the shifting macro-level landscape—focusing on climate finance, disclosure regimes, and regulatory frameworks—as context for the subsequent analysis of corporate reporting and strategic decision-making. The chapter then identifies eight decision-critical questions for business leaders, organised around three strategic pillars: data quality, risk management, and disclosure. These are intended to support commercially grounded responses to climate-related risks in an increasingly complex operating environment.

The Paris Consensus is unravelling

For much of the past decade, global efforts to address climate change have been shaped by what we term the 'Paris Consensus'—the belief that financial markets, corporate disclosures, and voluntary commitments are more effective at driving decarbonisation than direct state intervention. This approach assumes disclosures and commitments will lead to climate risks being properly priced, enabling capital markets to efficiently allocate resources towards sustainable investments and incentivising firms to decarbonise. Corporate purpose reporting—where firms articulate their social and environmental commitments—has evolved as part of this broader trend, relying on voluntary standards and investor expectations rather than government mandates.

This framework has shaped climate policy and financial regulation globally, influencing everything from corporate reporting requirements to central bank interventions and investment strategies. However, as climate risks intensify, questions about the effectiveness of these mechanisms have grown. While disclosure frameworks have expanded and sustainable finance has grown, their impact remains contested. Many argue that climate risks remain mispriced, corporate commitments lack accountability, and voluntary measures have failed to deliver meaningful emissions reductions. At the same time, political and market resistance to the Paris Consensus is increasing. This section examines the key pillars of this approach and evaluates their effectiveness.

Disclosure standards

A core assumption of the Paris Consensus is that climate-related financial disclosure enhances market efficiency by enabling investors to assess risk accurately and allocate capital accordingly. This also serves a broader financial stability function, ensuring that climate risks are incorporated into corporate risk management. The logic follows the principle that 'what gets measured gets managed.' By requiring firms to quantify and disclose climate risks, policymakers aim to create incentives for firms to mitigate those risks and strengthen market resilience.

Over the past decade, disclosure frameworks have expanded significantly, with the Task Force on Climate-related Financial Disclosures (TCFD) emerging as the dominant global standard. Established by the Financial Stability Board in 2015, the TCFD framework encourages firms to report on climate risks and opportunities across four areas: governance, strategy, risk management, and metrics and targets.

Building on the TCFD's foundation, regulators are moving towards requiring organisations—particularly publicly listed companies and financial institutions—to disclose climate-related financial information. This shift aims to enhance comparability across jurisdictions and ensure that firms provide consistent, decision-useful data to investors and stakeholders. The International Sustainability Standards Board (ISSB), established in 2021 by the IFRS Foundation, has developed the IFRS Sustainability Disclosure Standards (S2), integrating TCFD principles into a global baseline for corporate sustainability reporting (IFRS, 2023). Many jurisdictions, including the UK, Canada, Australia, and Singapore, are in the process of adopting these standards.

The European Union's Corporate Sustainability Reporting Directive (CSRD) goes further, requiring broader disclosure of environmental, social, and governance (ESG) factors. Unlike the ISSB's IFRS S2, which focuses on financial materiality, the CSRD mandates double materiality—requiring companies to disclose not only how sustainability issues impact financial performance but also how their operations affect the environment and society. While this results in more comprehensive reporting requirements in principle—across climate, biodiversity, water use, and governance—the EU has proposed significant adjustments to reduce the regulatory burden, effectively exempting four-fifths of previously targeted firms and postponing full implementation.

In the United States, the Securities and Exchange Commission proposed a rule mandating disclosure of material climate risks and their financial impacts by publicly listed firms. While the proposed rules incorporate TCFD principles, they diverge in certain areas, particularly regarding so-called 'Scope 3' emissions, which are only required under specific conditions. Scope 3 emissions include all indirect emissions from the reporting company's upstream and downstream activities.[2] The rule was challenged in court shortly after its adoption in March 2024, and enforcement remains on hold. In February 2025, the SEC announced it was pausing its defence of the rule.

Unlike climate disclosures that are primarily risk- or financially driven, purpose reporting focuses on a firm's broader obligations, responsibilities, and commitments to society and the environment—emphasising regeneration and long-term stakeholder value over shareholder control. Emerging standards for purpose reporting, such as PAS 808:2022, are designed to help organisations articulate how they deliver long-term value for stakeholders and society. PAS 808:2022 encourages firms to consider the systems, capitals, and relationships they depend on, and to report on how they avoid overexploitation and contribute to regeneration. This positions purpose not as an add-on but as a core organising principle for strategy, operations, and governance.

A key challenge is the difficulty of measuring and verifying the impact of corporate purpose. Unlike financial metrics, social and environmental outcomes are often complex, long-term, and difficult to attribute directly to a company's actions. This lack of clear metrics can hinder accountability. When discussing the adoption of purpose reporting standards, it is important to acknowledge the challenges in their implementation and the lack of standardisation, which makes comparisons difficult.

Disclosure frameworks have expanded in recent years, and there has been a corresponding rise in the number of firms reporting climate-related information, up from around 8% to over 20% in the EU and the UK, and from below 4% to over 12% in the USA between 2012 and 2022. However, many firms continue to under-report climate risks, limiting the utility of these disclosures for investors and regulators. Even among reporting firms, data quality and comparability remain inconsistent. Scope 3 emissions—often the largest component of a company's carbon footprint—are still voluntarily reported in many jurisdictions and frequently omitted or underestimated due to methodological inconsistencies.

Table 10.1 Summary of key climate and purpose disclosure frameworks.

Standard	*Institution*	*Focus*	*Scope/Approach*	*Notes*
TCFD	Financial Stability Board (2015)	Climate-related financial risk	Governance, strategy, risk management, metrics, and targets	Foundational for most current climate disclosure regimes
ISSB (IFRS S1 & S2)	IFRS Foundation/ISSB (2021–)	Sustainability and climate	Builds on TCFD, with global baseline for financial materiality	Adopted or planned by the UK, Canada, Australia, and Singapore
CSRD	European Commission (2022)	ESG with 'double materiality'	Includes climate, biodiversity, water, social factors, and governance	Broader than ISSB; full implementation delayed and narrowed
SEC Climate Rule (2024)	U.S. Securities and Exchange Comm.	Climate-related financial risk	Includes Scopes 1 and 2	Legal challenges and paused enforcement (as of February 2025)
PAS 808:2022	British Standards Institution	Purpose-led organisations	Emphasises obligations to stakeholders, systems health, and regeneration	Focused on how organisations serve broader societal purpose

Beyond coverage gaps, serious concerns persist regarding the credibility of reported climate data. Empirical studies highlight systematic weaknesses in emissions reporting, including inconsistent methodologies, strategic misreporting, and gaming of reported data. Aikman and colleagues (2023) highlight the potential in the current reporting protocol for firms to manipulate their emissions data to present a more favourable sustainability profile. The study, titled 'Emission Gaming,' takes activity data from three companies and finds stark differences between the largest and smallest feasible emissions estimates using the calculation methods and databases permitted under the GHG Protocol. On average across the three companies, the maximum estimate is between 4.6 and 6.7 times the minimum. If these firms are representative of small and medium-sized companies in the United Kingdom as a whole, the results suggest actual emissions from this sector could be materially higher than the 146 million tonnes of CO_2 equivalent (a standard unit for measuring the global warming potential of different greenhouse gases) currently reported. Such practices undermine the investor confidence, making it difficult to accurately assess transition risks and to price climate-related financial exposures.

Ultimately, the effectiveness of climate disclosure depends not only on transparency but also on enforcement. Without strong regulatory oversight, independent audits, and stricter accountability measures, disclosure regimes risk institutionalising greenwashing rather than driving genuine emissions reductions. These challenges highlight the limitations within the Paris Consensus framework. As the credibility of corporate climate reporting becomes increasingly critical, it is vital to strengthen these frameworks to address the identified problems and ensure they effectively support the low-carbon transition.

Climate finance

Climate finance mechanisms—including carbon pricing, green bonds, sustainability-linked instruments, and ESG investing[3]—are widely promoted as tools to accelerate decarbonisation. However, despite rapid expansion, their effectiveness remains contested. Concerns persist regarding their credibility—whether they genuinely reflect sustainable practices such as additionality—ensuring that the financed activities result in emissions reductions beyond what would have occurred without the investment, and impact—the actual environmental benefits achieved. Moreover, while these financial tools often align with a firm's stated purpose or mission, questions arise about whether they drive substantive change or merely serve as symbolic gestures.

Green bonds, for example, allow issuers to raise capital for sustainable projects, such as renewable energy infrastructure, energy efficiency, and climate adaptation. The Green Bond Principles, introduced in 2014, established voluntary best practices to enhance market transparency. These principles have since become the global benchmark for green bond issuance. In 2023, the EU introduced its Green Bond Standard, requiring at least 85% of proceeds to align with the EU Taxonomy for sustainable activities—a move intended to address greenwashing concerns.

Beyond green bonds, sustainability-linked loans (SLLs) and sustainability-linked bonds (SLBs) have gained traction. Unlike green bonds, which are tied to specific projects, these instruments link debt terms to ESG performance targets. In theory, this incentivises firms to improve sustainability performance through financial rewards, such as avoiding interest rate step-ups. However, many instruments feature weak, easily achievable targets, raising concerns about their credibility.

The green finance market has expanded substantially. By mid-2024, cumulative issuance of green, social, sustainability, and sustainability-linked bonds exceeded US$4.9 trillion, with green bonds alone surpassing US$3 trillion since 2006. Sovereign issuers have played a pivotal role, not only mobilising capital for national sustainability initiatives but also establishing benchmark pricing that facilitates private sector participation.

Despite growth in sustainable finance, concerns persist regarding green bond additionality—many finance projects that would have proceeded regardless. Some issuers have also included controversial activities, such as fossil fuel efficiency improvements, within their frameworks. Empirical studies highlight structural loopholes in sustainability-linked instruments, such as targets set close to maturity dates and call options allowing issuers to circumvent meaningful commitments (Lehman & Martins, 2023; Ul Haq & Doumbia, 2022).

Although green bonds were initially expected to benefit from a 'greenium'—a lower cost of capital relative to conventional bonds—empirical evidence suggests that any pricing advantage is marginal, typically in the range of 5–10 basis points. Investors remain sceptical about whether green bonds deliver additional emissions reductions, while regulators have generally eschewed preferential capital treatment for green assets. The difficulty in pricing climate risks—due to inconsistencies in disclosure, scenario modelling limitations (see Chapter 1), and regulatory fragmentation—has further hindered transition risk pricing in financial markets.

Reliance on voluntary corporate action

A core premise of the Paris Consensus is that companies will voluntarily reduce emissions if appropriately incentivised. Industry alliances such as GFANZ have sought to coordinate private-sector action, framing decarbonisation as a collective action problem requiring broad financial and corporate participation. Shareholder activism has also played a role, with investors pushing for climate alignment through engagement and resolutions.

However, recent research suggests that voluntary commitments have had little tangible impact on financial flows (Sastry et al., 2024). Many firms have retreated from ambitious targets amid political and economic pressures, and greenhushing—where companies under-report sustainability efforts to avoid scrutiny—is becoming more prevalent.[4] The withdrawal of financial institutions from GFANZ, along with hesitancy towards climate-aligned lending, underscores the fragility of voluntary initiatives. Without clear regulatory mandates, market-driven climate action is unlikely to drive systemic change.

A central role for central banks

Central banks and financial regulators have increasingly sought to integrate climate-related risks into their supervisory frameworks. Institutions such as the Bank of England have issued supervisory expectations outlining how banks and insurers should manage these risks, emphasising governance, risk management, and scenario analysis to safeguard financial stability. The Network for Greening the Financial System (NGFS) has played a central role in developing standardised climate scenarios to assess the financial implications of different climate pathways. These scenarios, which range from orderly and disorderly transitions to high-emissions 'hot-house' outcomes, form the basis of climate stress testing—exercises designed to evaluate how financial institutions might be affected by climate-related risks over short- and long-term horizons. While such tests have raised awareness, they have been criticised for underestimating extreme climate shocks, limiting their impact on regulatory decision-making.

Calls for central banks and regulators to adjust capital requirements based on climate risks— such as a 'green supporting factor' (reducing capital requirements for sustainable investments) or a 'brown penalising factor' (raising capital requirements for high-carbon assets)—have gained traction in some quarters. However, the European Banking Authority (EBA) has rejected these proposals, arguing that capital requirements should remain risk-based rather than serve as climate policy tools. Instead, regulators have focused on enhancing climate risk disclosure and scenario analysis within existing prudential frameworks.

Monetary policy has also faced pressure to incorporate climate considerations, particularly through 'green quantitative easing' (QE), where central banks tilt asset purchases towards sustainable investments. While the European Central Bank and the Bank of England have introduced modest climate tilts in their corporate bond purchase programmes, their overall impact remains limited as these programmes represent a small fraction of total central bank balance sheets. Moreover, with rising interest rates and the winding down of QE, central banks now have less scope to influence climate outcomes through asset purchases.

Despite the growing recognition of climate-related financial risks, a lack of political consensus has constrained central banks' ability to act decisively. As unelected institutions, central banks must operate within mandates set by governments, making climate policy a politically sensitive issue. Tucker (2018) argues that the legitimacy of independent institutions such as central banks depends on adherence to clearly defined mandates and democratic accountability. The withdrawal of the US Federal Reserve from the NGFS in 2025 underscores this political divide, reflecting broader scepticism towards regulatory intervention in climate policy. Without a clear and consistent political mandate, central banks face significant limitations in embedding climate risks into regulatory and monetary frameworks, leading to fragmented and uncertain approaches to climate finance.

The financial impacts of the climate crisis are intensifying

The financial consequences of the climate crisis are becoming increasingly severe, with significant implications for insurers, investors, and financial stability more broadly. Rising global temperatures and the increasing frequency and intensity of extreme weather events (see Chapter 2) have led to substantial economic losses, particularly within the insurance sector. At the same time, rising insurance premiums, market withdrawals, and trends in litigation signal a transformation in how financial markets and institutions are responding to climate risks.

One useful objective measure of the occurrence of extreme weather events in the United States and Canada is the Actuaries Climate Index (ACI) (American Academy of Actuaries and Canadian Institute of Actuaries, 2024). This composite measure tracks changes in the frequency of extreme high and low temperatures, heavy precipitation, drought, high winds, and sea level rise, relative to a 1961–1990 baseline. The index has increased steadily over time, with the 2024 reading at 2.1, indicating that climate anomalies are now occurring at over two standard deviations above historical norms.

In this section, we present evidence that the financial impacts of climate change appear to be intensifying. We first review developments in insurance markets, a key touch point for assessing the financial implications of climate risk. We then review trends in climate-related litigation.

Climate risk and insurance markets

The frequency and severity of billion-dollar insurance events have surged in recent decades. In 2023, a record 73 weather-related disasters exceeded $1 billion in damages worldwide, with total economic losses of $402 billion (World Meteorological Organization, 2023). Of this, insured losses accounted for approximately $140 billion, making it the third-costliest year for insurers on record. Severe convective storms, hurricanes, and wildfires have been major contributors to these losses. Hurricane Ida in 2021 alone caused $36 billion in insured damages, ranking among the most destructive hurricanes in US history. Similarly, the European floods of July 2021 resulted in $13 billion in insured losses, the largest flood-related insurance payout in European history. In response to growing climate-related losses, some reinsurers are reducing exposure to climate-exacerbated perils, transferring more risk to primary insurers and ultimately policyholders.

A direct financial consequence of escalating climate risks has been the sharp rise in insurance premiums and the withdrawal of coverage in high-risk areas (Swiss Re, 2021). In the United States, property insurance rates have surged, particularly in states exposed to hurricanes, wildfires, and flooding. Florida has emerged as a focal point, with average homeowners' insurance premiums nearly four times the national average. The state has also experienced a wave of insurer insolvencies, and more than two dozen either exiting the market or ceasing new policy issuance.

California has witnessed a similar retreat, particularly in wildfire-prone areas. The January 2025 Los Angeles wildfires resulted in estimated economic losses of between $250 billion and $275 billion, including the destruction of over 16,000 structures. Total property and capital losses ranged from $95 billion to $164 billion, yet insured losses were projected to be only $25 billion to $40 billion—highlighting a significant protection gap. Anticipating these risks, many insurers had already reduced their exposure, withdrawing coverage 18 to 24 months before the fires. With climate risks now being incorporated into premium calculations, policyholders face rapidly rising costs.

Similar trends are emerging globally. A study by the European Environment Agency (EEA) found that between 1980 and 2020, weather- and climate-related disasters across 32 EEA member states resulted in economic losses of €450 billion and €520 billion (in 2020 prices). However, only 25%–33% of these losses were insured, leaving a significant protection gap. The record-breaking floods in Germany, Belgium, and the Netherlands in 2021 led to a reassessment of flood risk pricing across the region. The UK's Association of British Insurers reported that storm-related claims in 2022 rose by 177% compared to the previous year, contributing to a 16% increase in average home insurance premiums.

In Australia, intensifying climate-related risks have made many regions commercially uninsurable. Insurers are either withdrawing from high-risk areas or setting premiums at prohibitively high levels. These shifts have significant financial and economic consequences beyond the insurance sector. In regions where coverage becomes unaffordable or unavailable, broader risks emerge, including depressed property values, constrained mortgage lending, and increased fiscal burdens on governments for post-disaster relief.

This trend is already evident in the United States, where the federally backed National Flood Insurance Program is over $20 billion in debt following repeated large-scale payouts. Similar pressures are emerging in Australia and Canada, where governments are considering expanded public–private insurance partnerships to address growing climate-related uninsurability.

The growth in climate-related litigation

Climate-related litigation has become an increasingly powerful tool for enforcing corporate and governmental accountability for climate change. The Sabin Center for Climate Change Law has tracked over 2,340 climate-related court cases globally as of mid-2023, more than double the number recorded in 2017—the date of the first documented cases in this database. Legal actions span a broad spectrum of issues, including corporate misrepresentation, government inaction, and financial sector responsibility for climate risks, with a surge in cases being brought against corporate 'climate-washing' in particular.

A landmark case illustrating corporate accountability is *Milieudefensie et al. v. Royal Dutch Shell*. In May 2021, a Dutch court ruled that Shell must reduce its global net carbon emissions by 45% by 2030, relative to 2019 levels. This ruling was unprecedented as it imposed a legally binding emissions reduction obligation on a private company. However, in November 2024, the Dutch Court of Appeal overturned the decision, concluding that while Shell has a responsibility to address climate change, setting specific reduction targets exceeds the jurisdiction of the courts. The reversal highlights the legal challenges in enforcing corporate climate commitments, though it has not deterred further climate litigation efforts against major emitters.

Financial institutions have also faced growing legal scrutiny. In Germany, Deutsche Bank's asset management arm, DWS, was investigated for allegedly misleading investors about the sustainability credentials of its funds. Regulators accused the firm of greenwashing—exaggerating the environmental impact of its investment products. In September 2023, DWS agreed to pay a $25 million settlement to the US Securities and Exchange Commission, with $19 million attributed to misstatements regarding ESG investment processes. Similarly, in the UK, HSBC was censured by the Advertising Standards Authority (ASA) in 2022 for misleading climate-related advertisements. The bank's campaign highlighted its green initiatives while failing to disclose its continued financing of fossil fuel projects. The ASA ruled that HSBC's omission of material information misled consumers, marking one of the first major rulings against a financial institution for greenwashing. These cases signal an increasing regulatory and legal focus on corporate sustainability claims, particularly in finance.

Pension scheme trustees are also facing increasing scrutiny over their fiduciary duties concerning climate-related investments. Trustees are legally obligated to act in the best financial interests of their beneficiaries, which encompasses considering long-term financial risks, including those posed by climate change (Association of Pension Lawyers (APL), 2024). A notable case highlighting this issue is *McGaughey and Davies v Universities Superannuation Scheme Ltd*, where beneficiaries alleged that the trustees failed to adequately consider climate risks by not divesting from fossil fuels. Although the Court of Appeal ultimately dismissed the claim on procedural grounds, the case underscores the growing legal challenges trustees may face if they neglect climate considerations in their investment decisions.

The rise of climate litigation has significant implications for financial institutions and insurers, as courts increase corporate liability for environmental impacts. Shareholder activism and investor pressure have intensified, with legal challenges prompting companies to disclose climate risks more transparently. Additionally, litigation is shaping regulatory policy, as governments facing legal action are compelled to implement stricter climate measures. The growing body of climate-related litigation also raises broader liability risks, particularly for financial institutions involved in financing carbon-intensive projects. As regulators tighten scrutiny on green finance claims, financial actors may face heightened legal exposure if sustainability-linked products or corporate disclosures are deemed misleading.

The preceding sections have explored structural shifts in climate finance, voluntary corporate commitments, and financial regulation, highlighting the growing complexity of sustainability-related decision-making. In this uncertain landscape, corporate leaders must assess not only compliance obligations but also their strategic positioning in an increasingly climate-conscious marketplace.

The following section outlines eight decision-critical questions to help business leaders navigate the evolving climate finance landscape, balancing financial, regulatory, and operational trade-offs. In this context, corporate purpose can be a valuable guide, helping to frame environmental and social commitments. However, it is crucial to recognise that purpose is not a substitute for robust risk management and financial planning. To be effective, purpose must be integrated with a clear understanding of climate-related financial risks and opportunities.

Decision-critical questions for business leaders

A primary concern for corporate decision-makers is the potential impairment of assets due to climate-related physical risks. The increasing frequency and severity of extreme weather events—such as flooding, wildfires, and subsidence—pose material threats to real estate values, supply chain continuity, and business operations. Assessing exposure to these risks and integrating resilience measures into corporate planning are critical to preserving asset value and ensuring business continuity.

Regulatory risks further complicate decision-making, as shifts in political leadership and policy direction create uncertainty. The rollback or reversal of climate policies at the local, national, regional, and international levels (discussed earlier in this chapter) may alter corporate obligations and reshape competitive dynamics.

Firms must remain agile, ensuring compliance with evolving regulatory requirements while positioning themselves to adapt to potential policy shifts.

Financial risks also play a crucial role in strategic planning, influencing operational costs and supply chain resilience. These include the potential introduction of climate-related tariffs, volatility in insurance coverage and premiums, and the increasing demand for skilled labour to support sustainability transitions. A failure to anticipate and address these risks could result in rising costs, business model disruptions, and diminished competitiveness in an increasingly climate-conscious market.

Drawing on insights from a decade of observation in climate finance, this section examines the key decision points, cost–benefit considerations, and trade-offs that corporate leaders must navigate. We structure this analysis around eight decision-critical questions that executives should consider, organised within three strategic pillars: data quality, risk management, and disclosure.

Data quality

Purpose-driven business is as much about data as it is about ethics. The introduction of the new IFRS Sustainability Disclosure Standards aims to integrate sustainability reporting with financial reporting by ensuring that material sustainability-related risks and opportunities are disclosed in a consistent and comparable manner. In this way, they provide a framework for disclosing how an entity's core mission and values influence its strategy and operations. The standards adopt the same definition of materiality as financial reporting: *Information is material if omitting, misstating or obscuring it could reasonably be expected to influence investor decisions*. This signals a fundamental shift—sustainability data and reporting are no longer peripheral to financial reporting but an essential component of it. As Richard Barker, a Director of the IFRS Sustainability Standards Board, noted, *it is all financial reporting* (From 'Accelerating Transition,' an event held at King's Business School in December 2024).

High-quality, decision-useful data are essential for both operational performance and asset valuation. Given the mounting evidence that climate change and biodiversity loss are impairing productive capacity and reducing asset values (as discussed earlier in this chapter), corporate leaders must ensure that sustainability-related risks and opportunities are effectively incorporated into financial planning. Three fundamental questions emerge for corporate leaders and purpose professionals:

Q1. How should businesses integrate climate-related risks into financial planning?

For purpose and sustainability data to meaningfully inform capital allocation and operational decision-making, it must be integrated with financial data in a robust, auditable manner. Business leaders require confidence that they are working with high-quality, verifiable purpose and sustainability metrics—data that meet the same financial-grade standards as traditional accounting measures. Without such integration, climate-related risks remain difficult to quantify, potentially leading to mispriced assets and inefficient capital allocation.

Some remain sceptical, arguing that such data are inherently less precise, often qualitative, and a distraction from the core business-as-usual objective of maximising shareholder returns. However, consistent and comparable GHG emissions data at the firm and sector levels are essential for investors to accurately price physical and transition risks. Reliable purpose and sustainability data underpin risk modelling and facilitate private capital mobilisation towards initiatives that mitigate or abate emissions.

A guiding principle for corporate leaders and investors should be: *Are these purpose and sustainability data audited to the same standard as financial data? If not, why not?* As regulatory frameworks evolve, the expectation for robust purpose reporting will continue to grow, reinforcing the need for accurate, decision-critical information in financial decision-making.

Q2. Which department should be responsible for purpose and sustainability data?

The question of responsibility for purpose and sustainability data remains a point of contention among business leaders. Some decision-makers view such reporting as an unnecessary complication that imposes additional costs without yielding tangible business benefits. From this perspective, if such data must be collected, they tend to be managed by a specialised team of purpose and sustainability experts, often operating on the periphery of the organisation. This approach treats purpose reporting as a compliance or public relations exercise rather than an integral part of financial and strategic planning.

Others argue that purpose and sustainability primarily serve reputational and branding objectives and should therefore be housed within the marketing or corporate communications function. This view reflects the historical role of corporate social responsibility (CSR) initiatives, which were often framed as voluntary commitments designed to enhance brand equity rather than as core business imperatives. However, as purpose reporting becomes increasingly tied to financial disclosures and investor expectations, this brand-building approach risks accusations of 'greenwashing' or 'purpose-washing,' where companies are perceived as insincere in their commitments, potentially damaging stakeholder trust, eroding investor confidence, and attracting regulatory scrutiny.

A growing consensus recognises the need to integrate purpose and sustainability reporting within core financial and risk management functions. The rigour, discipline, and professional standards associated with financial reporting provide a strong foundation for purpose and sustainability disclosures. Accountants and finance professionals bring expertise in data verification, materiality assessments, and compliance with regulatory frameworks—skills that are critical in ensuring the reliability and decision-usefulness of purpose and sustainability data. While additional training may be required to fully incorporate climate and ESG factors into financial reporting, the established principles of financial accounting, such as professional scepticism and the requirement to act in the public interest, make finance professionals well-suited to oversee purpose data management.

As purpose and sustainability disclosures become subject to the same scrutiny as financial reporting, firms must carefully consider how to institutionalise purpose and sustainability data management. The integration of purpose expertise with financial oversight offers the most effective path towards ensuring that climate-related risks and opportunities are captured with the same level of credibility and accountability as traditional financial metrics.

Q3. Can purpose and sustainability initiatives drive financial performance?

A common assumption among business leaders is that sustainability initiatives—such as reducing GHG emissions or water consumption—primarily represent a cost burden requiring substantial capital investment. This perception can act as a deterrent, particularly in an environment of regulatory uncertainty, where firms are hesitant to commit resources without clear policy direction.

However, an increasing body of evidence suggests that some sustainability initiatives can create financial value while simultaneously reducing environmental or social impact. For instance, the Food and Agriculture Organisation estimates that approximately 30% of food and drink produced globally is wasted (FAO, 2011). The hospitality industry alone accounts for around 12% of global food waste and loss. Case studies within this sector have demonstrated that targeted interventions to minimise food waste not only contribute to emissions reductions—particularly in Scope 3 emissions—but can also generate measurable financial returns (e.g. see the case study on Essendi in Box 10.1).

The success of such initiatives is often linked to strategic implementation, including employee training, performance incentives, and organisational support. For example, programmes that involve staff in identifying waste reduction strategies and offer recognition or rewards for successful initiatives can significantly enhance engagement and effectiveness. In the case of food service operations, effective programmes provide direct incentives to key decision-makers, such as head chefs, who oversee purchasing, preparation, and disposal processes. By aligning cost savings with budgetary flexibility, some firms have found that purpose and sustainability efforts can reinforce financial and operational performance. For instance, when catering departments achieve waste reduction targets, a portion of the savings may be reinvested in higher-quality ingredients, thereby aligning economic and environmental incentives. This approach not only motivates staff by directly linking their efforts to tangible benefits but also enhances the overall quality of service, creating a virtuous cycle of continuous improvement.

These findings highlight that sustainability initiatives should not be viewed solely as compliance-driven obligations but rather can be potential sources of competitive advantage. By reframing sustainability as an opportunity for cost-efficiency and innovation, businesses can unlock financial and environmental benefits while strengthening their resilience to regulatory and market shifts.

The case study of Essendi (Box 10.1) illustrates how reductions in GHG emissions can be achieved with modest capital expenditure, resulting in significant operational savings.

Box 10.1 Reducing food waste: Essendi

Essendi is a leading hotel owner and operator in the economy and mid-scale segments in Europe, with a portfolio of 566 owned or leased hotels across 22 countries, comprising more than 90,000 rooms as of March 2025. The company describes its purpose as to: 'create and foster places that thrive, continuously developing and innovating to generate long-term value for our assets, stakeholders, and ecosystem, with a constant commitment to responsibility and sustainability.' Like many large hospitality groups, Essendi faces operational challenges in improving efficiency while reducing its environmental impact.

One significant issue is food waste, which not only contributes to rising operational costs but also generates substantial GHG emissions, particularly methane (CH_4). Methane emissions from food waste in landfills are approximately 80 times more potent than carbon dioxide (CO_2) over a 15–20-year period, making food waste reduction a critical sustainability and cost-saving opportunity.

To address this challenge, Essendi implemented a targeted initiative to reduce food waste across a selection of its hotels. Aiming to onboard all hotels with significant food waste contributions between 2024 and 2026, the strategy combined smart investments with training programmes designed to optimise food purchasing, preparation, service, and disposal processes. In 2024, each participating hotel received an investment of approximately €7,000 in specialised equipment and training programmes for kitchen staff. The total expense for the first phase of the initiative—covering 50 hotels—amounted to €350,000.

The results of the programme were significant. On average, each hotel that participated in the initiative achieved annual savings of approximately €35,000. These savings were realised through more efficient food procurement, reduced waste disposal costs, and improved kitchen operations. At the same time, the initiative contributed to lowering the hotels' Scope 3 emissions by reducing the amount of organic waste sent to landfill. Essendi's experience highlights the financial and environmental benefits of sustainability-driven operational improvements. By integrating targeted investments with behavioural and process changes, the company not only enhanced its bottom line but also advanced its broader sustainability objectives. This case underscores the potential for well-designed corporate sustainability initiatives to create value both financially and environmentally.

Source: Essendi

Risk management

One approach to managing uncertainty is to prioritise short-term decision-making by building financial resilience in the form of capital reserves. This strategy seeks to ensure that businesses have the necessary resources to absorb future shocks.

However, this must be weighed against the opportunity costs of delaying investments in mitigation and adaptation measures. Deferring action until a crisis materialises may ultimately prove more costly, both financially and operationally.

The estimated costs of financing the global transition to net-zero emissions vary significantly. The International Energy Agency (IEA) projects that achieving the necessary clean energy transition to keep global warming well below 2°C will require approximately $4 trillion per year in clean energy investments by 2030 (International Energy Agency, 2021). The Intergovernmental Panel on Climate Change (IPCC) provides a broader range, estimating that between $3 trillion and $6 trillion annually is required to remain within 1.5°C–2°C warming pathways (IPCC, 2023). By contrast, the costs of inaction—or insufficient action—are estimated to be between $10 trillion and $20 trillion per year, representing an increase of between 200% and 500% relative to proactive mitigation and adaptation measures (see Chapter 1 for further discussion on the economic calculations of climate inaction). These figures underscore the economic imperative of strategic climate risk management.

Business leaders must carefully evaluate the trade-offs between allocating resources towards capital reserves and investing in climate mitigation and adaptation initiatives. While the financial benefits of resilience investments are well-documented, uncertainty surrounding future regulatory requirements complicates strategic planning. In jurisdictions where regulatory frameworks are evolving rapidly—such as the European Union—firms that proactively integrate climate risk management may benefit from preferential financing and improved investor perception. Conversely, in regions experiencing regulatory pushback or uncertainty, firms may hesitate to commit capital without clear policy signals, weighing the risks of premature action against potential future compliance costs.

However, decision-making in this space is further complicated by the fact that climate change presents a classic collective action problem. While businesses, governments, and individuals would all benefit from coordinated efforts to mitigate climate risks, the costs of unilateral action can be significant. In the absence of a coordinated regulatory framework, firms that invest in climate resilience may face competitive disadvantages relative to peers who opt to delay action or free ride on the mitigation efforts of others. This dynamic creates a dilemma for corporate leaders, who must balance the immediate financial costs of proactive climate strategies against the long-term benefits of reducing risk exposure.

The following questions may help executives assess their firm's approach to climate resilience and risk management:

Q4. Over what time horizon should businesses assess climate-related decisions?

The intensifying impacts of climate change require corporate leaders to navigate multiple, often conflicting, timescales when making strategic decisions. Cutting emissions and implementing purpose initiatives frequently demand significant upfront investments over the short to medium term, typically within the next three to five years. However, the benefits—such as reduced exposure to climate-related disasters and greater resilience to their impacts—may take considerably longer to materialise, and in some cases, may remain uncertain.

Investment decisions, particularly those concerning capital expenditure and resource allocation, must therefore be evaluated across varying temporal horizons. These can be categorised into distinct timescales: financial time, which operates on quarterly and annual reporting cycles; organisational time, typically spanning three to five years in many Western economies; tenurial time, which reflects executive leadership cycles—for instance, the average tenure of an S&P 500 CEO in the United States is now between four and eight years, depending on the sector; political time, typically over a term lasting four to five years with a maximum of two terms; legal time, which can extend from two to twenty years plus; cultural time, generally considered to span a generation—approximately twenty-five years; and environmental time, where the effects of GHG emissions persist in the atmosphere for hundreds of years, depending on the specific gas and its global warming potential.

Balancing these overlapping time frames is essential for aligning corporate strategy with long-term purpose-related goals. The case study of SH LLC illustrates how a well-designed sustainability initiative can deliver financial and operational benefits across multiple timescales—enhancing customer experience, reducing GHG emissions, and lowering operating costs.

Box 10.2 Optimising warehousing for sustainability and efficiency: SH LLC

SH LLC, an interiors and homeware retailer, faced a challenge in scaling its warehousing capacity to meet growing demand in the United States. To support rising sales, the company needed to expand its storage and distribution network by 30%. However, doing so in a way that minimised operational inefficiencies and reduced environmental impact required a departure from the business-as-usual approach.

While SH LLC had sustainability goals such as reducing carbon emissions and promoting responsible consumption, the decision to optimise warehousing was driven by strategic considerations, aligning environmental benefits with operational efficiency.

Under the existing logistics model, imported goods arrived at the main US ports before being transported to a central distribution centre in Chicago. This structure had been established when most of its customer demand was concentrated on the East Coast. However, a closer examination of customer data revealed a shift over time in demand to the West Coast. This shift in demand called into question the efficiency of continuing to rely solely on Chicago as the primary distribution centre.

Rather than expanding its Chicago facility, SH LLC opted to open a second warehouse in California. This strategic decision allowed the company to accommodate growth while significantly improving logistics efficiency and reducing its carbon footprint. By distributing inventory closer to West Coast customers, SH LLC improved delivery times by an average of seven days, enhancing customer experience while simultaneously cutting transport-related CO_2 emissions by almost 30% in 2024 compared to expanding solely in Chicago. The new approach also yielded material operational savings, with further efficiencies expected in the years ahead.

Looking at a ten-year horizon, this initiative is projected to reduce transport-related CO_2 emissions by approximately 4.6 million kilograms, a 40% reduction compared to operating exclusively from Chicago.

This case highlights how sustainability considerations can align with commercial objectives to drive both financial and environmental benefits. It also illustrates how a clearly stated corporate purpose can be operationalised through decisions that support long-term value creation across environmental, customer, and financial dimensions. By integrating sustainability analysis into strategic decision-making, SH LLC was able to optimise its logistics network in a way that reduced costs, enhanced customer satisfaction, and contributed to corporate emissions reduction targets.

Source: SH LLC

Q5. Should businesses explore green financing, carbon credits, or climate insurance to hedge against risks?

As discussed earlier, initial expectations that sustainable finance would provide preferential incentives, such as a lower cost of capital (the 'greenium'), have largely not materialised. Despite this, firms must weigh the absence of direct financial benefits against the increasing likelihood that capital markets will demand robust climate risk management as a prerequisite for access to investment.

A key requirement under emerging purpose and sustainability accounting standards, including the IFRS sustainability disclosure framework, is the obligation to report a credible net-zero plan. However, net zero and carbon neutrality are often conflated despite fundamental differences. Net zero requires firms to prioritise direct emissions reductions, using offsets only for residual emissions that cannot be mitigated. This may involve investment in carbon capture technologies, such as pre-combustion carbon capture or direct air capture. In contrast, carbon neutrality typically relies on purchasing carbon credits—permits the emission of greenhouse gases in exchange for financing external mitigation initiatives, such as forestry projects or improved cookstove distribution. While these mechanisms may reduce emissions elsewhere, they do not necessarily require the firm itself to decarbonise, raising concerns about their credibility.

Beyond financing mechanisms, firms may consider climate insurance as a tool for hedging against risk. Climate insurance enables businesses to transfer the financial burden of climate-related losses to insurers in exchange for a premium. Policies may cover direct asset damage, business interruption, or parametric insurance, in which payouts are triggered by predefined climate conditions, such as temperature, wind speed, or rainfall levels.

As discussed earlier, climate risk is increasingly shaping real estate values and insurance markets, with insurers reassessing exposure in high-risk regions. As premiums rise or coverage becomes unavailable, firms must evaluate how the loss of insurability affects asset values, business continuity, and long-term investment decisions. Ultimately, businesses must balance the costs of adaptation and mitigation against the risks of physical damage, rising insurance costs, and restricted access to capital. As financial markets and insurers continue integrating climate risk into pricing and underwriting decisions, firms that fail to adapt may face not only environmental risks but also declining valuations and financial constraints.

Disclosure

The catalyst for the IFRS to develop Sustainability Accounting Standards came from the International Organization of Securities Commissions (IOSCO), the global network of securities regulators in 132 jurisdictions. IOSCO recognised the growing financial risks posed by climate change and the critical need for standardised reporting to enhance transparency and comparability in corporate disclosures. In 2024, one month after their publication, the IFRS standards were unanimously endorsed by IOSCO members, marking a significant step towards global regulatory alignment on climate-related financial reporting.

Climate policy increasingly emphasises private sector funding as a key mechanism for financing the transition to net zero. The effective allocation of capital towards initiatives and technologies that mitigate and abate GHG emissions depends on the ability to accurately price both physical and transition risks. The IFRS Sustainability Disclosure Standards aim to facilitate this by requiring firms to disclose GHG emissions data—or, where necessary, the highest-quality estimates available—in a manner that is consistent and comparable across firms, sectors, and financial reporting frameworks. This integration of sustainability metrics with financial data is essential to improving risk modelling and investment decision-making. Without such standardisation, there is a risk that greenwashing could contaminate financial data, misleading investors and distorting capital allocation.

New sustainability disclosure regulations and accounting standards seek to counteract greenwashing by ensuring that corporate sustainability claims are substantiated by verifiable, decision-useful data. Regulators argue that misleading disclosures undermine market integrity, distort efficient capital allocation, and impede the broader goal of sustainable economic transition. By embedding climate-related risks and opportunities within financial reporting, the IFRS standards aim to enhance transparency, improve investor confidence, and promote more informed decision-making across financial markets. In parallel, emerging frameworks such as PAS 808:2022 offer guidance for purpose-led organisations, encouraging them to disclose how their activities contribute to long-term value creation for stakeholders and the systems they depend on.

Q6. Should businesses actively communicate their sustainability stance and purpose or adopt a low-profile approach?

Corporate leaders must navigate a complex and evolving regulatory and reputational landscape when determining how to position their businesses on purpose and sustainability. They face three broad strategic choices:

(i) actively advocating for the adoption of sustainability accounting standards and regulations, including through purpose-led organisational change
(ii) resisting such developments by supporting regulatory rollbacks and reversals
(iii) adapting to purpose and sustainability expectations while maintaining a low-profile approach to avoid potential political and reputational backlash

For multinational corporations, the decision is further complicated by the extraterritorial reach of certain climate-related regulations. Policies such as the European Union's Carbon Border Adjustment Mechanism (CBAM) are designed to apply to companies irrespective of their country of incorporation. Even firms headquartered in jurisdictions with weaker domestic climate policies may find themselves subject to regulatory pressures in global markets, reinforcing the importance of proactive compliance with sustainability disclosure requirements.

Many corporate leaders are opting for a pragmatic approach, continuing to implement sustainability initiatives without publicising their efforts extensively. This strategy recognises the evolving expectations of customers, employees, and business partners while also mitigating exposure to regulatory and financial risks (Conference Board, 2023). By systematically collecting and analysing purpose and sustainability performance data, businesses can uncover new opportunities to enhance financial performance while maintaining voluntary disclosure without high-profile commitments that might attract external scrutiny. While some argue that public articulation of purpose is a core element of being purpose-driven, others see strategic discretion as a way to preserve flexibility in politically or commercially sensitive environments.

Moreover, purpose and sustainability disclosure can offer strategic advantages in negotiations with investors and creditors. Firms that can demonstrate verifiable reductions in GHG emissions through financial-grade, auditable data may improve their bargaining position when accessing capital. A key consideration for financial institutions is their Scope 3 emissions, commonly referred to as 'financed emissions'—the emissions associated with the capital they provide to investees or debtors. Companies that successfully reduce their emissions contribute to lowering the financed emissions of their lenders and investors, potentially making them more attractive counterparties in financial transactions.

Table 10.2 Estimated impact of a 10% rise in financed GHG emissions for cost of equity.

Bank	*Original cost of equity (%)*	*New cost of equity (%)*
HSBC	10.2	10.4
Lloyds Banking Group	13.3	13.4
NatWest Group	12.7	12.8
Barclays	15.3	15.5
Banco Santander	14.3	14.4
Standard Chartered	12.8	12.9

Notes: Each bank's original cost of equity is calculated by taking the average of estimates of the daily returns of each bank's equity in relation to the daily returns of the MSCI EAFE Index and the FTSE All-Share Index for selected banks from January 2021 to August 2023 (column 1). Column 2 shows the new CoE following the increase in each bank's reported Scope 3 emissions. These estimates reflect the reported Scope 3 emissions of each bank.

Preliminary empirical research supports this financial rationale for sustainability disclosure. A study conducted by researchers at King's College London found that banks reporting a 10% increase in CO_2 emissions might face an increase in their cost of equity of up to 20 basis points (Table 10.2). The findings also suggest that banks reporting lower CO_2e emissions might expect to benefit from a corresponding reduction in their cost of equity. While these estimated impacts might appear modest in absolute terms, they would represent substantial financial benefits for a bank with billions of dollars in equity capital.

Ultimately, transparency in purpose and sustainability reporting can serve as a tool for corporate leaders to manage and mitigate financial risk while ensuring access to the broadest possible funding pool on the most favourable terms. As regulatory frameworks and investor expectations continue to evolve, businesses that can credibly demonstrate their sustainability performance may secure a competitive advantage in capital markets.

Q7. Should businesses voluntarily adhere to purpose and sustainability standards even if regulations are reversed?

Corporate decision-makers must assess the implications of potential regulatory reversals for their business strategy and financial positioning. While some may view regulatory rollbacks as an opportunity to scale back purpose and sustainability commitments, others may anticipate that financial markets will continue to demand robust climate risk management as a condition for capital access. The decision to maintain voluntary adherence to purpose and sustainability standards should be weighed against both immediate compliance costs and potential long-term shifts in investor expectations and regulatory landscapes.

A focus on energy efficiency and supply chain sustainability can deliver cost savings and operational efficiencies, irrespective of regulatory changes (see case studies). Investments in climate-adaptive infrastructure and risk management practices may also help mitigate disruptions associated with extreme weather events, enhancing business resilience. However, the financial justification for such investments depends on sectoral dynamics, geographic exposure to climate risks, and investor expectations, all of which vary across firms and industries.

Voluntary purpose or sustainability (or ESG) commitments may also influence relationships with key stakeholders, including customers, employees, and investors. Some businesses may see alignment with purpose or sustainability expectations as a way to strengthen brand reputation and maintain trust, while others may determine that a lower-profile approach is preferable in an increasingly polarised landscape. Investors, even in jurisdictions where regulatory oversight remains limited, are progressively integrating sustainability metrics into decision-making, but the extent to which this influences capital allocation decisions remains subject to debate.

Ultimately, there is no single correct approach. Firms must weigh the potential costs and benefits of voluntary adherence to purpose and sustainability standards in light of their financial strategy, regulatory environment, and stakeholder priorities. While some may choose to maintain strong purpose-related and/or sustainability-related commitments to future-proof their business, others may prioritise flexibility in response to evolving market and policy conditions.

Q8. Should businesses integrate sustainability—and their broader purpose—in their supply chains?

Supply chain emissions, categorised as Scope 3 under GHG accounting frameworks, frequently account for over 80% of a company's total carbon footprint. However, requiring suppliers to report emissions has generated debate, with firms citing the complexity, cost, and resourcing burdens involved. Some organisations are addressing this challenge by embedding carbon pricing into procurement decisions, using emissions data as a criterion for supplier selection. While such initiatives incentivise reductions across the value chain, they also introduce potential trade-offs, including administrative complexity and potential supplier resistance. Business leaders must consider the costs and benefits of supply chain decarbonisation, balancing competitive dynamics with long-term purpose and sustainability commitments.

One approach to integrating sustainability and purpose into supply chain management is through innovative procurement practices. The Sustainable Procurement Pledge, an international non-profit organisation comprising procurement professionals, academics, and industry practitioners, has proposed a model that directly embeds carbon pricing into supplier selection criteria. By incorporating emissions data into commercial evaluation processes, this method effectively internalises carbon costs that have historically been external to financial decision-making.

Under this approach, supplier bids are assessed based not only on the price of goods and services but also on the associated carbon cost. This is calculated using the supplier's reported total GHG emissions linked to the provision of goods or services, multiplied by the prevailing price of carbon under the European Emissions Trading System. Suppliers that do not disclose their emissions are assigned a default carbon cost based on the industry average plus a 50% penalty, significantly reducing their competitiveness in the bidding process.

For businesses adopting such procurement models, potential advantages include incentivising emissions reductions across the value chain, improving the quality of emissions data for financial reporting, and enhancing sustainability performance in the eyes of investors and other stakeholders. However, firms must also consider the potential downsides, including increased administrative complexity and possible resistance from suppliers operating in jurisdictions with weaker reporting requirements. The effectiveness of such measures will depend on industry norms, regulatory developments, and the broader willingness of suppliers to participate in carbon disclosure initiatives.

Summary

While political cycles and regulatory frameworks may fluctuate, climate risks and market shifts represent persistent, long-term challenges for businesses. Whether and how firms choose to integrate sustainability and purpose into their corporate strategy will depend on their assessment of long-term cost savings, market positioning, and investor expectations, balanced against short-term regulatory and operational considerations. Corporate decision-makers can evaluate these trade-offs using the eight decision-critical questions we proposed, summarised in Table 10.3.

Conclusion

The evolving landscape of climate finance, purpose, corporate sustainability, and financial regulation presents a complex and uncertain environment for business leaders. The unravelling of the Paris Consensus, the shifting role of financial markets, and the intensifying financial impacts of climate change highlight the challenge of integrating long-term climate risk into business strategy. As firms navigate these uncertainties, their approaches to purpose and sustainability can

Table 10.3 Eight decision-critical questions for corporate leaders to consider.

Data quality
Q1. How should businesses integrate climate-related risks into financial planning?
Q2. Which department should be responsible for purpose and sustainability data?
Q3. Can purpose and sustainability initiatives drive financial performance?
Risk management
Q4. Over what time horizon should businesses assess climate-related decisions?
Q5. Should businesses explore green financing, carbon credits, or climate insurance to hedge against risks?
Disclosure
Q6. Should businesses actively communicate their sustainability stance and purpose or adopt a low-profile approach?
Q7. Should businesses voluntarily adhere to purpose and sustainability standards even if regulations are reversed?
Q8. Should businesses integrate sustainability—and their broader purpose—in their supply chains?

be broadly categorised into three alternative perspectives: compliance-driven, conviction-driven, and business-driven decision-making.

- *Compliance-driven* actors focus primarily on meeting regulatory requirements and investor expectations, responding to evolving disclosure mandates and financial reporting standards. For these firms, sustainability efforts are often shaped by the need to maintain market access and mitigate potential legal or reputational risks.
- *Conviction-driven* firms, by contrast, view purpose as a core strategic imperative, embedding long-term environmental considerations into corporate governance and capital allocation. Their engagement with purpose and sustainability is not solely reactive but reflects a broader commitment to climate action as a fundamental business principle.
- *Business-driven* approaches, meanwhile, focus on identifying financial opportunities within purpose and sustainability, leveraging operational efficiencies, cost reductions, and shifting consumer and investor preferences to create competitive advantages.

Regardless of the motivation, the financial implications of climate risk are becoming increasingly difficult to ignore. The physical impacts of climate change that we reviewed in the first chapters of the book are reshaping risk management practices, while regulatory and investor scrutiny is evolving in ways that will likely continue to influence access to capital. Firms that effectively integrate sustainability into their corporate strategy or that become purpose-led may enhance resilience, secure long-term financial stability, and maintain stakeholder trust. However, the precise balance between regulatory compliance, financial performance, and environmental responsibility will differ across industries and individual firms.

As this chapter has demonstrated, there is no singular path forward. Climate finance remains a contested space, voluntary corporate initiatives are facing mounting challenges, and financial markets continue to struggle with the pricing of climate risk. Business leaders must weigh these competing dynamics, recognising that their decisions today will have lasting implications for their organisations' financial and operational future. The eight decision-critical questions outlined in this chapter provide a framework for navigating this uncertainty, offering a structured approach to balancing risk, compliance, and opportunity in an era of increasing climate-related financial pressures.

Notes

1 Greenwashing or purpose-washing refers to the practice of misrepresenting an organisation's sustainability performance by selectively presenting information, exaggerating environmental claims, or omitting material details. This can take the form of overstating progress on emissions reductions, misclassifying business activities as sustainable, or selectively reporting favourable sustainability metrics while ignoring adverse impacts. While greenwashing can sometimes result from misinterpretation or poor data quality, it is often a deliberate attempt to align corporate narratives with investor and consumer expectations.

2 Scope 1 emissions are direct greenhouse gas emissions from sources owned or controlled by the company, such as emissions from company vehicles or on-site fuel combustion. Scope 2 emissions are indirect emissions resulting from the generation of purchased electricity, steam, heating, or cooling consumed by the company. Upstream Scope 3 emissions include those from purchased goods and services, capital goods, fuel- and energy-related activities not included in Scope 1 or 2, upstream transportation and distribution, waste generated in operations, business travel, employee commuting, and upstream leased assets. Downstream Scope 3 emissions include those from transportation and distribution, processing of sold products, use of sold products, end-of-life treatment of sold products, leased assets, franchises, and investments.

3 Carbon pricing assigns a cost to greenhouse gas emissions, incentivising emitters to reduce their carbon footprint. Green bonds are fixed-income securities designed to fund projects with positive environmental benefits, such as renewable energy initiatives. Sustainability-linked instruments tie financial terms to the issuer's achievement of predefined ESG objectives. ESG investing involves incorporating ESG factors into investment decisions, aiming to promote sustainable and ethical practices.

4 For instance, a recent report found that most surveyed companies, across nearly every major sector of the economy, are actively decreasing their climate communications ('greenhushing'). See https://www.southpole.com/news/survey-finds-most-companies-going-quiet-on-green-goals.

Further reading

Carney, M. (2021). *Values: Building a better world for all.* Penguin Random House.

Lepere, M. (2023). Green investing: The global system for rating companies' ethical credentials is meaningless. *The Conversation.* https://theconversation.com/green-investing-the-global-system-for-rating-companies-ethical-credentials-is-meaningless-179724

World Economic Forum. (2024). *The cost of inaction: A CEO guide to navigating climate risk.* https://www.weforum.org/publications/the-cost-of-inaction-a-ceo-guide-to-navigating-climate-risk

References

Aikman, D, Dong, Y, Drellias, E, Havaldar, S, Lepere, M, & Nilsson, M. (2023). *Emissions gaming? A gap in the GHG Protocol may be facilitating gaming in accounting of GHG emissions.* King's Business School Impact Paper No. 1. https://www.kcl.ac.uk/business/assets/pdf/research-papers/kbs-research-impact-paper-1-emissions-gaming.pdf

American Academy of Actuaries & Canadian Institute of Actuaries. (2024). *Actuaries climate index: 2024 update.* https://actuariesclimateindex.org/home/

Association of Pension Lawyers (APL). (2024). *Written evidence from the Association of Pension Lawyers (APL): Legislative and parliamentary sub-committee, investment and DC sub-committee (FYD0013).* https://committees.parliament.uk/writtenevidence/128407/pdf/

Conference Board. (2023). *Sustainability practices 2023: Trends in corporate sustainability reporting.* Conference Board. https://www.conference-board.org/

Food and Agriculture Organisation. (2011). *Global food losses and food waste.* https://www.fao.org/4/mb060e/mb060e.pdf

Intergovernmental Panel on Climate Change. (2023). *AR6 synthesis report: Climate change 2023.* IPCC. https://www.ipcc.ch/report/ar6/syr/

International Energy Agency. (2021). *Net zero by 2050: A roadmap for the global energy sector.* IEA. https://www.iea.org/reports/net-zero-by-2050

International Financial Reporting Standards. (2023). *IFRS sustainability disclosure standards.* https://www.ifrs.org/issued-standards/ifrs-sustainability-standards-navigator/ifrs-s2-climate-related-disclosures/

Lehmann, A., & Martins, C. (2023, March). The potential of sovereign sustainability-linked bonds in the drive for net zero. *Bruegel Policy Brief*. https://www.bruegel.org/system/files/2023-03/PB%2007%202023_2.pdf

Sastry, P., Verner, E., & Marques-Ibanez, D. (2024). *Business as usual: Bank climate commitments, lending, and engagement*. European Central Bank Working Paper Series 2921. https://www.ecb.europa.eu/pub/pdf/scpwps/ecb.wp2921~603e225101.en.pdf

Swiss Re Institute. (2021). *The economics of climate change: No action not an option*. Swiss Re. https://www.swissre.com/institute/research/topics-and-risk-dialogues/climate-and-natural-catastrophe-risk/the-economics-of-climate-change.html

Tucker, P. (2018). *Unelected power: The quest for legitimacy in central banking and the regulatory state*. Princeton University Press.

Ul Haq, I., & Doumbia, D. (2022). *Structural loopholes in sustainability-linked bonds*. World Bank Policy Research Working Paper Series. https://doi.org/10.2139/ssrn.4099829

World Meteorological Organization. (2023). *State of the global climate 2023*. WMO. https://library.wmo.int/doc_num.php?explnum_id=11650

Index

Note: Page numbers in *italics* indicate figures, **bold** indicate tables in the text, and references following "n" refer to notes.

accountability 113, 115–116, 145, 147, 191–192; corporate 67, 70, 210; Danone 176; democratic 208
Actuaries Climate Index (ACI) 208
adaptation 7, *53*, 67; investment in 69; mitigation efforts and 24, **41**, 49, 68, 216
aerosols 30, 38
affective MS capabilities 103
Agentforce (AI-powered platform) 178
Alliance of Small Island States (AOSIS) 83
alumni relations *172*; in purpose-led organisations 177–179
analytics: digital HRM 177, 178; digitalisation 171; machine learning and 178; onboarding phase 174; people analytics 175; people management 181; purpose-aligned L&D 175; sustainable HRM 180–181
Arrhenius, Svante *4*, 28
artificial intelligence (AI) 171; EU Artificial Intelligence Act 181; gamification approaches 174; Salesforce 178; Unilever 175; use by purpose-led organisations 180; use in recruitment and onboarding 174
Asia Energy (mining firm) 94
Australia 210; climate-related risks 209; Great Barrier Reef (GBR) 54; modelling future climate 37; news on claims of climate change 33; study on domestic violence 59

biodiversity: credit 68, 69; crises 1; loss/decline 7, 17, 43, 53–56, 70, 120, 126, 154, 212; UK's Biodiversity Net Gain market 69
British Standards Institute (BSI) 13, 113
business assets, threats to 63–65
business-as-usual (BAU) 14, 111, 114, 118–122, 135, 138, 217
business development *4*
business-driven approaches 224
business leaders: data quality 212–215, **223**; decision-critical questions for 211–223, **223**; disclosure 219–223, **223**; risk management 215–219, **223**
business operations 17, 60, 63–65, 211

Callendar, Guy S. *4*, 29
capability-seeking 140
capitalism 50; shareholder 9, 10, 18, 91, 92, 95, 97, 105; stakeholder 13, 17, 18, 91–95, 97, 100–101, 105
carbon cycle 5, 6, 37; carbon reservoirs *38*
carbon dioxide (CO_2) 3, *4*, 5–6, 17, 24, 27, *28*, 29, 37, 39, 221; comparative ability to warm atmosphere **36**; compared with methane emissions 215; past variations in 25–26; rise from pre-industrial times 35; transport-related emissions 218
carbon emission 6, 7, 15, 18, 27, 29, 31, 37, 39, 40, 45, 116, 126, 210, 217–218; absorbed by natural carbon cycle 37; carbon taxes on 87; global warming and *42*; global warming by 2100 *42*; impact on Earth temperature 27; societal and economic causes of 45
carbon isotopes 6, 25
carbon leakage 66
Carbon Neutral Cities Alliance (CNCA) 88
carbon pricing 87, 206, 222, 225n3
carbon sink 5, 6, 27, 29, 37
cascading risks 66–67
central banks 207–208

China: bilateral green investments 82; blackouts in 65; domestic emissions reduction targets 84; GHG emissions 78, 80, 82; technology due to mass market production in 85
chlorofluorocarbons (CFCs) 31; comparative ability to warm atmosphere **36**
Cisco 145
Climate Action 100+ (CA100+) 67
climate change, history of 27–34; delay in recognising 29–30; environmental social movement 30–31, 34; impacts of 32–33; media and 33–34; science 28–29
climate change, scientific evidence for *4*, 5–8, 34–45; modelling future climate 37–40; tipping points 40–43
climate finance 79; limitations of 201; mechanisms 206–207; platforms 86; structural shifts in 211; transnational capital 75; under UNFCCC 81–82
climate justice 7, 62, 67, 92
climate policymaking 76–80, 201; corporate purpose reporting 203; private sector funding 219; regulatory intervention in 208
climate-related litigation 20, 208, 210–211
climate risk 202–203, 221; biodiversity 1, 7, 17, 43, 53–56, 68–70; conflict and migration 55, 60–61; into decision-making 20; disclosure 203, 204, 208, 211; financial implications 68, 207, 219, 224; food security 17, 34, 49, 56–58; geographic exposure to 222; infrastructure 63–65, **64**; human health 5, 13, 49, 53, 58–60; insurance markets and 65, 209–210, 219; labour markets 17, 61, 62; mitigation 216; real estate 219; supply chains 11, 34, 49, 56, 60, 62, 66, 67, 70; vulnerabilities and social inequalities 50
'Colleagues for Life' initiative 179
commoditisation 56
common but differentiated responsibility 7, 62, 75, 76, 84
common good HRM 171
communication, purpose-led 187–199; aspiration story 196–197, *198*; best practice guidelines 193–194; character story 197, *198*; corporate purpose 187–190; organisation's purpose 196; power of communication 199; profiling exercise 190–198; purpose communication strategy 191; purpose narrative 196, *198*; purpose statements 193–194; stakeholder story 197–198, *198*
communication model canvas *189*
communities 9, 10, 60, 94, 98; coastal 54; diverse 142; impacts on 62–63; marginalised 18, 67, 101; migration 61; vulnerable 7, 56
compliance-driven actors 224
compliance-seeking, in firm preparedness 140
Conference of the Parties (COP) 67, 75–77, 85; COP15 77, 78; COP21 7, 77; COP26 *42*, 77, 81, 202; COP28 67, 77, 81; COP29 11
constituency groups 85
conviction-driven firms 224
cooling effects 38–39
Copenhagen Accord (COP15) 77; failure of 79
corporate greenwashing 67
corporate purpose 138, 154, 166, 187–190, 202, 204; decision-makers 155, 160, 161; non-financial aspects of 153; non-financial elements of 160; organisation design 163; performance management system 164; related to culture 144; reporting 203
corporate social responsibility (CSR) 13–14, 92, 121–122, 136
cost–benefit analysis 154
Coupled Model Intercomparison Projects (CMIP6) 39
Covid-19 33
creating shared value (CSV) 121–122
cross-border risks 66–67
cross-functional leadership 181–182

Danone 136, 176
data quality, business leaders 212–215, 223
decarbonisation 7, 14, 67, 88, 201, 203, 206–207, 222
decision-critical questions 201, 202, 211–223
decision-making, purpose-led 76, 126, 128, 152–168, 213–214; change in 167–168; cognitive challenges faced by decision-makers 155; context and purpose 157–158; contextual layers influencing *156*; culture 165–167, *167*; data availability 154–155; formal

organisation 161–167; individual decision-makers 159–161; leadership 162–163, *167*; methods used in 155; organisational factors influencing *167*; organisation design 163–164, *167*; performance management 164–165, *167*; problems faced by organisations 154–155; processes in 163; resolving issues of competing commitments 156–167; self decisions 158–159; social organisation 161–167
deforestation 2, 6, 24, 27, 39, 51, **52**, 67
Deloitte 179; Deloitte Alumni Network 179
Deloitte Access Economics 54
disclosure, business leaders 219–223, **223**
disclosure frameworks 201, 203, 204, **205**, 218
disclosure standards 203–206
discourse analysis 15
Doughnut Economics (Raworth, Kate) 32, 126
drought 1, 7, 35, 43, 44, 53, 55, 56, 60, 62, 66, 208
duty-based purpose 12

economic growth 8, 10, 32, 39, 50, *51*, 67, 69; degrowth 17; *Limits to Growth* 30
economics 118; of climate change 32–33; development 100; neoliberal 11
ecosystem-based adaptation (EbA) 68–70, 208
ecosystems 68–69; biodiversity loss 54; impact of increased weather events on 50; marine 34; natural resources and 53–55; oil spills damage 95; terrestrial 34
egalitarian approach 82
emissions: scope 3; emissions 204
employee involvement 171, 176, 177, 178
employee life cycle: sustainable HRM in *172*, 173–179
enabling stakeholders 125; *see also* stakeholders
Enhanced Transparency Framework (ETF) 86
entertainment MS capabilities 104
environmental and social governance (ESG) 67, 70, 120–121, 137, 138, 143, 146, 204, 210, 222
Environmental Kuznets curve (EKC) 50, *51*
environmental social movement 30–31, 34
equity MS capabilities 103
Essendi 215
European Union (EU) 77, 81, 83; Artificial Intelligence Act 181; Green Bond Standard 206
exit management 177–179
European Union's Emissions Trading Scheme 78
Extinction Rebellion 34
extreme weather 2, 7, 33, 35, 55; climate-related 65; disaster losses 50; disruptions 49; disruptions associated 221; food production 44; manufacturing processes and global trade 63; migration 61; in United States and Canada 208

fake news 3, 33
farmers 26–27, 37, 56, 86, 155, 197–198
financial development *4*
financial disclosure 65, 203, 213
financial risks 65, 201, 208, 211, 220, 221; of climate change 201, 219; climate risk 68; firms role climate change mitigation 66; strategic planning 212
flooding 2, 7, 44, 49, 58, 63, 209, 211
food security 126; agricultural productivity 17; climate change impact 34; organisations in mitigating 56–58
Forbes 145
Forest Principles 31
fossil fuels 2, 3, 5, 6, 10, 15, 25, 29, 39, 84, 87; climate risks 211; efficiency 206; replaced by renewable energy 44; reserves *38*; scientific consensus 11; subsidies 8
France 68, 81, 87, 113, 136; Champagne industry 57
Fridays for Future movement 34
Future Fit Plan (Unilever) 175

G77 + China 83
General Circulation Models (GCMs) 37
General Data Protection Regulation (GDPR) 181
Germany 76, 81, 85, 87, 209, 210
GHGs *see* greenhouse gases
glaciers 2, 34, 35, 43
Glasgow Climate Pact 42, 77
Glasgow Financial Alliance for Net Zero (GFANZ) 202
global mean temperature (GMT) 30
Global North 7, 62, 106
Global Reporting Initiative (GRI) 146
Global South 7, 62, 106
Global Stocktake (GST) 86

global warming 3, 5, 6, 17, 38–39, 43; clean energy transition 216; combat 11, 18; effect on biodiversity loss 17; externality 49; harmful effects of 53; impacts of 7; marginalised stakeholders (MSs) 92; relationship to carbon emission *42*; science of 29; theory of 30
goal-based purpose 12
Great Barrier Reef Foundation 54
green growth 50, 67–70
greenhouse gases (GHGs) 3, 5–6, 25, *25*, 35, 76, 84, 225n2; carbon credits 218; carbon pricing 225n3; comparative ability to warm atmosphere **36**; economic growth and *51*; Holocene GHGs 26; reducing emissions 32, 195; sector contributions to **52**; temperature 26, *26*; warming effects of 38
green HRM 171
greenhushing 202, 207, 225n4
greenwashing 67, 93, 152, 202, 206, 210, 213, 219, 224n1
gross domestic product (GDP) 11, 15, 17, 27, 32, 50, *51*, 62–63, 82

habitat loss 54
healthiness MS capabilities 102
heat-related impacts on productivity 59
high ambition coalition (HAC) 83
horizontal approach 82
How to Save Our Planet (Maslin, Mark) 17
human health 5, 13, 53, 58–59; organisations' role in 60
human resource management (HRM), sustainable 170–180; balancing purpose and profit via 179–180; consistency and personalisation in 180; in employee life cycle *172*, 173–179; HR departments 164; technologies and analytics 180–181

ice sheets 2, 24, 25, 26, 34–35, 42–44
IFRS Sustainability Disclosure Standards 212
individual decision-makers 159–161; *see also* decision-makers
induced conflict, organisations' role 62
industrial revolution 6, 25–27, 30, 37, 39, 44
inequality (SSP4) **41**, 45, 105, 126, 143
influential stakeholders 125
infrastructure risk, organisations' role in 66
innovation 14, 67–70, 88, 111, 197
insurance: climate 63, 218–219; financial 65; markets 209–210; premiums 208, 209, 212; resilience 69
integrity MS capabilities 102–103
interconnected risks 66–67
intergovernmental organizations (IGOs) 85
Intergovernmental Panel for Climate Change (IPCC) 7, 24, 25, 30–31, 50, 216; Representative Concentration Pathways (RCPs) report 40, **41**; Sixth Assessment Report (AR6) 39–40; Special Report on Global Warming of 1.5°C 34; Special Report on the Ocean and Cryosphere 34; Stern Report 32
International Displacement Monitoring Centre 61
International Energy Agency (IEA) 216
International Sustainability Standards Board (ISSB) 65, 147, 203, **205**
investment: adaptation 10, 59, 214; carbon capture technologies 66, 218; decisions 217; environmental impact of 210; ESG 210, 225n3; in nature-based solutions 68–70; into renewable energy 70; risk modelling 219; in sustainable innovation 67
IPCC *see* Intergovernmental Panel for Climate Change
ISO37000 **123–124**

Japan 81; Joint Crediting Mechanism 84
just transition 68, 202

Keeling, Charles *4*, 29; Keeling curve *4*, 29
Kyoto Protocol *4*, 76–80, 82, 86

labour markets 17, 61
land clearing 27
land use 2, 6, 24, 37, 39–40
LDC Expert Group (LEG) 86
leader/leadership 139, 194; ambitions 112; cross-functional 181–182; cycles 217; development 111; dilemma 136; empowering 141–143; political 211; purpose-driven 141–142, 148; structural disruption 135; *see also* business leaders; decision-making, purpose-led
learning and development (L&D) purpose-aligned 174–175
least developed countries (LDCs) 82, 83, 86
legally binding agreement 79
legitimacy 95–98, 99, 104–105, 146–147, 162, 208
Local Communities and Indigenous Peoples Platform (LCIPP) 86

macro-level challenges 14–15
managing people, purpose-led organisation 170–182; challenges and recommendations for 179–182; sustainable HRM in the employee life cycle *172*, 173–179; technologies and analytics in 181
marginalised stakeholder capitalism (MSC) 95, 96, 100–101, 106
marginalised stakeholders (MSs) 92–96; capabilities 101–105, **102**; in stakeholder–shareholder relationship 97–100
Marrakech Partnership (MPGCA) 86
Mauna Loa observatory 27; atmospheric carbon dioxide *28*
meta-purpose 111, 118, 120, 126–128
methane (CH_4) 5–6, 24, 215; atmospheric emissions contributors 45n1; comparative ability to warm atmosphere **36**
migration 34, 35, 54, 61
mission 144, 190; organisational 115–116, 174–175, 179, 180, 181; Salesforce 178
mitigation 6, 7, 11, 24, 34, *53*, 66, 216, 218–219; climate change 137; proactive 216; public–private partnerships 68; socio-economic **41**, 49; SSPs 40
Montreal Protocol 31
MSs *see* marginalised stakeholders

Nairobi Work Programme (NWP) 86
National Adaptation Plans (NAPs) 86
nationally determined contribution (NDC) 7–8, 45n2, 80
nations, lack of trust 79
natural resources 50, 51, 53–55, 69, 93, 96, 103
nature-based solutions (NbS) 68–70
NDC partnership 86
negotiating groups 80–85
negotiation: failure in 79; sustainability disclosure 220; UNFCCC 83–85
neoliberal economy 8–11, 14, 92
nitrous oxide (N_2O) 5–6, 24; comparative ability to warm atmosphere **36**
non-governmental organisations (NGOs) 18, 30, 76, 85, 93, 94, 99, 101, 104–106, 140, 143
non-national government organisations 85–88
Nordhaus, William: future discount rates 32

observer groups 85–88
OECD 59, 81
Ofwat 10
onboarding 173–174
organisational challenges 15–16
organisational change 68, 70, 88, 134, 135, 138–140, 220
organisational culture 19, 103, 138, 142, 165, 172, 181
organisational identity 114, **123**
organisational mission 115–116
organisational objective 114
organisational preparedness 140
organisational purpose 110–129, 188, 196; business-as-usual organisations 118–122; corporate social responsibility (CSR) 121–122; creating shared value (CSV) 121–122; defining 114, 122; environmental and social governance (ESG) 120–121; history 114; measuring and reporting on purpose 202–206, 212–214, 217, 219, 221, 223, 224; non-negotiable commitment to no harm 126; organising principle 114; parameters 126–127; purpose-driven organisation 122–129, **123–124**; purpose statement 125–128; purpose summary 128–129; relationship between vision, mission, and strategy 115–116; serving the good of others 116–118; strategic decision-making and 114–115; strategic organisational parameters 127; sustainability 116–118; values and 119–120
organisational responsibility 49–70; climate-driven resource conflicts and migration 60–61; factors of climate change 50–53; financial risks, disclosure, and insurance 65; food security 56–58; human health 58–59; impacts on communities 62–63; infrastructure 63–65; interconnected, cascading, cross-border risks 66–67; to mitigate climate change impacts on human health 60; to mitigate climate change impacts on infrastructure and financial risk 66; to mitigate climate change induced conflict 62; natural resources, impact on 53–55; opportunities for innovation and green growth 67–70
organisational strategy 115–116
organisational transformation 134–135
organisational values 116, 119–120, 122, **123**, 127, 176, 191
organisational vision 115–116

organisation design 162, 163–164, *167*
organisation structure 163, 166
organising principle 114, 129, 204
ozone (O_3) 24, 28, 31

palaeoclimate/past climatic change 30, 35
Paris Agreement 7–8, 77–84, 86, 88, 137, 202–211, **205**, 223
Paris Consensus 203, 206, 207, 223
PAS 808 20, 113, 120, 122, **123–124**, 126, 128, 129, 204, **205**, 219
passive resistance 141
Patagonia 12, 174
perfluoro methane PCF-14 (CF_4) **36**
performance management 146, 156, 164–165, *167*, 171, 175–177, 182
personalisation, in sustainable HRM 180
person-organisation (P-O) fit theory 173
physiological MS capabilities 101–102
political and material control MS capabilities 104
polluter pays principle 82
policymaking, international climate 76–80
power of communication 199
primary beneficiary stakeholders 125
profiling exercise 190–198
Prosperity Without Growth (Jackson) 32
public discontent 79
public protests 79
purpose-aligned L&D 174–175
purpose communication strategy 191
purpose narrative 196, *198*
purpose reporting 145, 203, 204, 213
purpose statements 125–128, 193–194; best practice guidelines 193–194
purpose summary 128–129
purpose-washing 15, 112–113, 129, 135, 202, 213, 224n.1

Rabobank 196–197
radiative forcing 40
Rana Plaza 99
rationality MS capabilities 103
Reagan, Ronald: opinion on climate change 9
recognition MS capabilities 103
recruitment 173–174
reporting standards 20, 204, 224
Representative Concentration Pathways (RCPs) report 40, **41**
resistance: active resistance 141
resources: allocation 105, 180, 203, 216; migration 60–61; natural 53–55, 69, 93, 96, 103; scarce 2, 60; transformation of 117
responsibility *see* organisational responsibility
Revelle, Roger:on oceans as a carbon sink *4*, 29
rewards management, purpose-driven 175–176
Rio Declaration on Environment and Development 31
Rio Earth Summit 31, 44, 77
risk management 203, 207, 213, 215–219, 221, **223**

Salesforce 178
scenario 17, 24, 30, 39–40, **41**, *42*, 43–44, 95, 207
Scientific American 29
shareholder capitalism 9, 10, 18, 91–92, 97
sharing MS capabilities 103–104
Shell 95
SH LLC 217–218
Small Island Developing States (SIDS) 2, 44, 84
small to medium-sized companies (SMEs) 15–16, 62
socially responsible HRM 171
social organisation, purpose-led 161–167
societal tipping points 42
stakeholder capitalisms 13, 17, 18, 92–95, 97, 105; *see also* marginalised stakeholder capitalism
stakeholders: affected stakeholders 125
Stern, Nicholas 32, 49
Stern Report *4*, 32
strategic decision-making 114–115, 201
subsidies 8, 85
sulphur hexa-fluoride (SF6) **36**
supply chains 34, 49, 56, 60, 127, 155; decarbonisation 222–223; disruptions 70; resilience 212; sustainability 222–223; threats to 63–65
sustainability 2, 3, 8, 10, 13, **41**, 76, 77, 110, 116–118, 176, 179, 181, 192; challenges 112; data 213–214; disclosures 213–214, 219, 220, 221; ecosystem protection and 70; global transition to 87; initiatives 206, 214, 215, 220; integration in businesses 222–223; meaningfulness and 116–118; objectives 175; oriented HRM framework 171; performance 206, 224n1; promotion 86; related decision-making 211; related risks and opportunities 212; standards 221; thought-leader 194–195; warehousing for 217–218

Sustainability Accounting Standards Board (SASB) 146
Sustainable Development Goals 13, 17, 31, 34, 126, 137, 171
sustainable finance 137, 201, 203, 206, 218
sustainable HRM *see* human resource management (HRM), sustainable

Task Force on Climate-Related Financial Disclosures (TCFD) 65
Technology Mechanism (TEC & CTCN) 86
Thames Water 10
Thatcher, Margaret: opinion on climate change 9–10, 31, 92
Thinking Fast and Slow (Kahneman) 161
Thunberg, Greta 34
ticking the box approach 144
tipping points 40–43, 70
Traditional Owner groups 54
transformation, purpose-driven 18, 19, 112, 134–147; approaching resistance in 140–141; case for 135–138; changing to 138–147; Cisco's approach to culture and purpose 145; embedding purpose in culture 143–144; five stages of organisational preparedness 140; identifying starting point 139–140; leader's dilemma 136; leadership 141–142; measuring and reporting performance 144–147; stakeholders 142–143; sustainable finance 137
Trump, Donald: opinion on climate change 11
Trump Administration 8, 82, 85
Tyndall, John: demonstration of greenhouse effect 28

ultimate beneficiary stakeholders 125
Ultra Low Emission Zone (ULEZ) 87
Umbrella Group 83
Unilever 175
United Kingdom (UK) 10, 30, 33, 82, 84, 160, 204; British Academy 12; Climate Risk Assessment 1–2; Environment Act 66; Extinction Rebellion 34; setting biodiversity credit market with France 68; small and medium-sized companies 205; Stern Report 32; tree planting targets 67; water crisis in 2
United Nations (UN) 12, 85; Conference on Environment and Development (UNCED) 31; Convention on Biological Diversity 31; Convention to Combat Desertification 31; Framework Convention on Climate Change (*see* United Nations Framework Convention on Climate Change); Millennium Development Goals 31; Sustainable Development Goals (SDGs) 13, 17, 31, 34, 126, 137, 171
United Nations Framework Convention on Climate Change (UNFCCC) 7, 18, 31, 75, 76, 81–82; coordination platforms 86; COP28 (Dubai, UAE) 77; Copenhagen Accord (COP15) 77; Glasgow Climate Pact (COP26) 77; Kyoto Protocol 77–80; negotiations 83; observer organisations 85; Paris Agreement (COP21) 7–8, 77–80, 82–84, 86, 88, 137, 202–211, **205**, 223; Rio Earth Summit 31, 77
United States (US) 81, 202; climate-related damages 50; National Oceanic and Atmospheric Administration 1; oil and gas companies 33; property insurance rates 209; purpose of corporation 134; Securities and Exchange Commission 204; Senate Committee on Energy and Natural Resources 30
University of Cambridge Institute for Sustainability Leadership (CISL) 139
Urban Heat Island effect (UHI) 59
urgency 93, 97–98, 138, 155

values *see* organisational values
value-seeking 140
vertical approach 82
Vienna Convention for the Protection of the Ozone Layer, 1985 31
voluntary corporate action 207
vulnerability 44, 50, 56, 60, 63, 66, 84, 161

water shortages 60, 62
Water UK 10
wellbeing 13, 118, 124, 126, 153
wellbeing economy 111
World Economic Forum 11–12

9781032898957